T0418951

Building Materials, Health and Indoor Air Quality

In *Building Materials, Health and Indoor Air Quality: Volume 2* Tom Woolley uses new research to continue to advocate for limiting the use of hazardous materials in construction and raise awareness of the links between pollutants found in building materials, poor indoor air quality and health problems. Chapters in this volume reinforce previous arguments and present new ones covering:

- Further evidence of the health impacts of hazardous emissions from materials
- Hazardous materials to be avoided and why
- Fire and smoke toxicity – the Lakanal House and Grenfell Tower legacy
- Sub-standard retrofits leading to damp and mould in previously sound houses
- A critical review of recent reports from UK Government and others on air quality and health problems including policy changes on flame retardants
- Growing evidence of cancer risks and the failure of cancer research organisations to address these issues
- Critical review of recent climate change and zero carbon policies and a discussion on whether extreme energy efficiency is a good thing

This book asks some important and, for some, uncomfortable questions, but in doing so it brings to light important areas for research and provides much needed guidance for architects, engineers, construction professionals, students and researchers on hazardous materials and how to reduce their use and design and build healthier buildings for all occupants.

Tom Woolley started his career as a practising Architect in Scotland and London. He moved into research and teaching, working in Glasgow, Hull and Belfast where he was Professor of Architecture at Queen's University. He is the author of numerous books including Natural Building Techniques, Thermal Insulation Materials for Building Applications and was co-editor of the award-winning Green Building Digest. He has continued to lead seminars and hands-on workshops about natural building construction throughout the world.

Building Materials, Health and Indoor Air Quality

Volume 2

Tom Woolley

LONDON AND NEW YORK

Designed cover image: Ian Sidaway

First published 2024
by Routledge
4 Park Square, Milton Park, Abingdon, Oxon, OX14 4RN

and by Routledge
605 Third Avenue, New York, NY 10158

Routledge is an imprint of the Taylor & Francis Group, an informa business

British Library Cataloguing-in-Publication Data
A catalogue record for this book is available from the British Library

ISBN: 978-0-367-65387-3 (hbk)
ISBN: 978-0-367-64669-1 (pbk)
ISBN: 978-1-003-12922-6 (ebk)

DOI: 10.1201/9781003129226

Typeset in Times New Roman
by Newgen Publishing UK

To Hannah, without whom the book wouldn’t have been published

Contents

Acknowledgements

Special thanks to Jessica Pereira, Jamie Page and Rachel Bevan
Lisa Ponzoni
Hannah Bevan Woolley
Oliver Bevan Woolley
Tim Robinson
Kristine Reilly Blake
Outi Ilvonen
Anna Maria Scutaru
Alison Pooley
Hugh Dunn
Tim Pye
Lulie Anderson
Julia Bennett
Laura Fatah
Marion Roberts

1 Introduction

The aim of this book is to inform the reader of the dangers of chemical emissions from building materials and other related health hazards in the home and workplace. The use of chemicals in composite building products has increased significantly in recent years and this has led to higher levels of hazardous emissions inside our homes, schools and offices. A further aim is to suggest how it is possible to avoid many of these chemicals, so it is possible to have much healthier buildings. We are subject to pollution from a variety of sources, from the water we drink, to the air we breathe, from food and every aspect of life. The "chemical cocktail" problem reflects a growing dependence on petrochemical, plastic and synthetic aspects to lives, throughout the planet, and the resulting pollution is one of the most significant causes of climate change, loss of habitats and wildlife, as well as health and development problems.

There are some who play down, or even dispute the problem of chemical emissions from building materials, so a key reason for this book is to set out the significant evidence about the issue, and also provide a major critique of the lack of proper policies from a wide range of organisations, to ensure we have healthy homes and buildings. University researchers, Government departments and a range of voluntary and campaigning organisations have their heads in the sand about the use of chemicals in buildings, either promoting the use of synthetic, plastic and chemical products, or diverting attention to issues such as cooking, cleaning and wood burning stoves. There has been greater emphasis on external air and traffic pollution than indoor air, which is undoubtedly a problem. There is no doubt that there are health benefits from reducing emissions from vehicles, and the issue of low emissions zones in UK cities, is constantly in the news, but this is not an excuse to devote massive resources to studying traffic pollution while ignoring indoor chemical pollution.

The book also explores the way in which modern buildings are built and older buildings renovated that can lead to poor indoor environments, particularly the massive growth in the problem of damp and mould, which has even led to deaths and serious illness. The problem of mould is blamed on occupants for creating too much moisture and failing to ventilate their homes sufficiently. Even when mould is recognised, as a cause of ill health and even death, the reports ignore the way in which the buildings and building materials have caused condensation and mould growth, and guidance about dealing with the problem is inadequate.

The book builds on the first volume *Building Materials, Health and Indoor Air Quality*, published in 2017, but this time exploring evidence and information in more detail, but also commenting on the inadequate science and policy guidance from "experts" and Government advisors. Much of the research on chemical emissions drawn on, is from the USA, Europe and around the world while the UK and Ireland is somehow out of touch with this extensive scientific body of knowledge.

DOI: 10.1201/9781003129226-1

We inhabit a world that is increasingly dominated by synthetic chemicals and they can have a serious negative effect on our health as well as the wider environment. Many architects, builders, officials and members of the public welcome the use of these chemicals as they are convinced that they are much better than what was used in the past, that they are cheaper and more robust and generally better for all of us. It's not easy to understand how this belief developed and why so many people cling to it tenaciously, even when confronted with evidence of the damage chemicals can cause, but the evidence of the dangers of chemical emissions, which have increased dramatically in the last few decades, is overwhelming. Many people have suggested that the Grenfell disaster, with 72 people killed in the fire on June 14 2017, would be a game changer and be a wake-up call to the construction industry about the dangers of chemical and plastic materials, but nothing could be further from the truth. Since the fire, despite the scandalous evidence that emerged in the Inquiry, sales of dangerous plastic materials have increased, though with a small shift to materials claiming to be less flammable.

The growth of chemical use throughout the world was first identified by Rachel Carson in her 1962 book Silent Spring. Initially focussed on pesticides she realised the dangers of wider pollution and the connections with cancer. She herself got breast cancer. Despite her warnings and others that followed, there exist many internet sites today arguing the benefits of chemicals in everyday life, unsurprisingly issued by petrochemical companies like Shell **(1)**.

Despite dire warnings of global warming and environmental catastrophe, which is taken as a given here, the world is still reliant on massive overuse of fossil fuels, not just for heating and transport but also to make the millions of chemicals we have come to depend on for everyday life. UNEP estimates the global chemical industry to have exceeded $5 trillion but this may have doubled by 2030 **(2)**. The dependence on plastic, and the well-publicised problem of plastic waste, continues to increase and the construction industry, with some limited concession to recycling, makes bigger and bigger profits from sales of synthetic chemical-based products.

Medical researchers have linked the fact that men in the Western world produce half as much sperm as they did 40 years ago, due to exposure to toxic chemicals. Studies also show that exposure to toxic chemicals results in girls entering puberty earlier, increasing the risk of getting breast cancer later in life. There are many organisations drawing attention to these problems, but a good place to start to get an overview is through Chemsec, based in Sweden. Not only do they document the chemical problems but they also work with industry to try and develop alternatives **(3)**.

Taking a stand against the powerful world of chemicals requires great bravery and persistence. Rachel Carson faced many personal attacks, not just for her scientific work but as a woman, and she was inevitably labelled a communist. It is worth mentioning a few other brave heroes who have fought against the chemical industry and their pollution and damage to health. Erin Brokovitch is well known, as portrayed by actor Julia Roberts, but she is also still working as a paralegal fighting current cases. She was recently summoned into action following the chemical spillages from a recent train crash in Palestine Ohio in the USA **(4)**.

Hollywood created a film of lawyer Roger Billot's fight against DuPont on behalf of hundreds of thousands of people affected by polluted water, with the actor, Mark Ruffalo (playing Billot) becoming an environmental campaigner himself **(5)**. Having Hollywood to promote such issues is an advantage, but the US legal system, allowing class action lawsuits, and the role of the, far from perfect, US Environmental Protection Agency (EPA), still allows much greater access to information about chemical production, emissions and prosecutions. Very little similar access or information is available in the UK where the Environment Agency has been slow to take action against foreign-owned water companies that have been accused of pouring sewage into UK's rivers and seas for instance **(6)**.

There have been other important campaigners in the USA such as Arlene Blum, a formidable mountaineer, but also an environmental health scientist and chemist who has done much to raise the issue of flame retardants and PFAS **(7) (8)**.

American Architect Hal Levin, an early pioneer of the building ecology movement, became a leading figure in the field of indoor air quality for several decades, giving its study a status barely achieved in the UK even today **(9)**. Chares Weschler, a chemist, has drawn attention to the high levels of chemicals in building materials **(10)**. Despite the status of Blum, Levin and Weschler, none were too busy or self-important to have found the time to send this author the occasional friendly and helpful email in the past.

The Guardian published a bizarre list of the top 100 environmental campaigners, including some who actively promote the use of chemicals, but hardly anyone who can be said to have addressed the issues in this book apart from Colin Ward **(11)**.

There have been important figures in the UK but most of them are not well known. What tends to distinguish them is their holistic understand of issues. Richard Douthwaite, an alternative economist **(12)**, had a wide ranging a holistic approach, which included the built environment, but the organisation he helped to found FEASTA in Ireland now concentrates on sustainable economics **(13)**. Jamie Page, a chemist who founded the cancer prevention society in 2001, a UK-based charity, working to reduce the incidence of cancer and other diseases raises awareness of the connection between diseases and exposure to chemical substances **(14)**. A couple of UK academics, Professor Anna Stec in Preston **(15)**, and Dr. Michelle Bellingham in Glasgow **(16)**, have done vital work identifying chemical dangers to health and have had to withstand personal attacks. Eddie Daffarn a resident at Grenfell in London deserves a mention as he too was dismissed as a trouble maker before the Grenfell fire, but he would undoubtedly say that all the people who died, and those who survived at Grenfell, were heroes and the issues which they have raised have still not been properly addressed. This book is partly inspired by them **(17)**.

Professor Stephen Holgate in Southampton is a hero because, even though part of the UK medical establishment, he was the first to draw attention to indoor air quality in a ground-breaking report for the Royal College of Paediatrics and child health **(18)**. There are many others who have taken a stand against the big chemical and building materials companies and this does not come without certain risks.

Sadly, there are many other academics and scientists in the UK who are happy to go along with accepting funding from the chemical industry, thus compromising the independence or credibility of their research, as explained by Wright et al in a review of how UK Universities address these issues

> Research partnerships with business are often promoted as a panacea, contributing much needed resources while supporting important research into societal challenges. The UK government and its public research funding agency, UK Research and Innovation (UKRI), are both enthusiasts; UKRI has encouraged industry collaboration to help restore the UK economy in the wake of covid-19 and promoted the co-production of health innovations. The publicly funded UK Prevention Research Partnership, an initiative to reduce non-communicable diseases, similarly encourages "engage[ment] with industry in the business of prevention." **(19)**

Taking funding from industry has led to a compliant culture among some academics working on issues of building and air quality. They avoid tricky areas that might bring them into conflict with potential funders and promote solutions that are influenced by vested interests. The much-applauded Salford Energy House at Salford University has built test houses with Barratt Developments, Bellway homes, and Saint Gobain and while some useful scientific research has been published, this remains largely uncritical of current building and retrofit practices, as promoted by those companies and others. It was not possible to find any investigation of chemical emissions or indoor air quality work in the test house reports **(20)**.

A range of overlapping committees and advisory bodies decide who gets funding and should be asked to speak at meetings and conferences in the UK. Their reach is extraordinary and extends into the voluntary sector as well as government and wider. Personal experience of being frozen out..or to use a modern term, "cancelled" has become so much more evident to this author, since the 2017 book appeared, though sometimes it is very useful as it is possible to identify opponents and friends. and also spot biased academic and scientific bias, some of which will appear in later chapters.

One of the most worrying aspects of this problem is the obsession of Government, and many professionals and others in the construction industry, prioritising energy efficiency over everything else. This can be seen as partly responsible for the increased use of chemicals in buildings and the health problems that are the subject of this book. Since the 1960s to 80s, when discussions began about sustainable construction, and reducing operational energy, the leading figures in the energy efficiency movement took very clear decisions to focus on making buildings with synthetic construction materials without concern for their potential negative consequences.

When we published the "Green Building Digest" and the Green Building Handbook in 1997 (Woolley T. et al Green Building Handbook Taylor and Francis 1997) we set out four holistic principles of Green Building going beyond the narrow focus on energy efficiency **(21)**.

The construction industry, even today, has barely caught up with this holistic approach.

(a) Reducing Energy in Use
(b) Minimising External Pollution and Environmental Damage
(c) Reducing Embodied Energy and Resource Depletion
(d) Minimising Internal Pollution and Damage to Health

Twenty-Five or more years later there has been some discussion in the mainstream construction industry about the problem of embodied energy, for instance, but this is still somewhat tokenistic. The fossil fuel energy used to create low energy buildings still remains high, which is contradictory but energy efficiency advocates will continue to promote the false claim that payback is just a matter of a few years, ignoring the energy used to build the buildings.

> Embodied carbon is changing the way we look at sustainable construction. For example, some materials used in energy-efficient homes, such as plastics, allow for lower operational carbon but have a high embodied carbon. Considering the embodied carbon of the home during construction may lead to choosing a different material with lower resulting emissions overall. **(22)**

Policy documents from a very wide range of organisations promoting zero carbon or net zero buildings, rarely if ever discuss how these buildings will be built. The UK Green Building Council report on a framework to achieve net zero published in 2019 for instance does not mention the word insulation one single time other than to acknowledge the assistance of Kingspan Insulation. It is hard to understand how zero carbon buildings can be created without using insulation! Their recent web site claims to have moved on somewhat from the 2019 report but the UKGBC cannot risk being critical of the big multi-national chemical companies as they rely on them for membership and funding **(23)**.

Some thirty years ago some of the big players in the chemical insulation market barely existed. Kingspan for instance set up by Eugene Murtagh, making trailers in Ireland, was only floated on the stock Exchange in 1989. Kingspan's well known insulation products such as "Kooltherm" only emerged onto the market in 1996 **(24)**.

It is not uncommon for architects to dismiss alternative healthy, chemical free building materials, low embodied energy products, as too risky because they say they are "innovative and untried", and yet hempcrete buildings were being built in 1996 at the same time that Kooltherm was being launched, just as innovative and untried! Celotex the other material which featured in the Grenfell enquiry began as a fibreboard company in the 1920s using asbestos. Celotex began to produce PIR foam board much earlier than Kingspan in 1978 **(25)**. Celotex was bought by the Saint Gobain group in 2012.

These two companies, are referred to here as well-known examples of how recently the vast expansion of chemical based building materials became part of normal construction. In Chapter 3 a large range of chemicals and their possible health effects are listed that have found their way into almost every component of modern buildings, with their accompanying emissions.

The industry became convinced that this was the way to create energy efficiency buildings, as the many chemical insulation materials were presented as much more efficient than what had been used before. In reality many of the new chemical-based products are not that much more efficient and very little research has been done to assess the performance of the resulting "low energy" buildings. The concept of post occupancy evaluation has been largely ignored but where studies have been done, the performance is often much worse than predicted **(26)**.

The government funded Zero Carbon hub, established in 2008, did such a great job exploring the failure of buildings to achieve the hoped-for energy efficiency, referred to as the "performance gap," that its Government funding was withdrawn in 2016 and much of their work buried out of sight. Fortunately, an archive of their work remains and is worth inspection **(27) (28)**. Research by academics into the performance gap problem tended to focus on poor building skills, practice and management ignoring the possibility that the new chemical foam materials could have been underperforming. Some of the research on this problem was not surprisingly funded by insulation suppliers **(29)**.

One way in which manufacturing industry has been able to promote chemical materials is to influence the development of standards, certification, testing and building regulations, as was revealed during the Grenfell inquiry. Peter Apps in his chapter "We will be quids in" points out that industry capitalised on a culture of deregulation which allowed materials and products to be used without proper testing and controls **(30)**.

Bisby, one of the official expert advisors to the Grenfell inquiry, has set out at length how the standards and regulation culture endorsed by Government and the industry was based on an outcome rather than prescription approach

> Over many decades, the enabling of "innovation" in support of economic gains appears to have been central to regulation and practice… Hackitt's recommended approach (after the Grenfell fire) reiterated the same ideology that underpins the Building Act 1984 – in her recommendations she sought a regulatory system that was "truly outcomes-based (rather than based on prescriptive rules and complex guidance)".
>
> Central to the issues that I have identified in this report is the process of technological innovation. The hazards of new materials, products, and systems appear to have been insufficiently understood or, where they were understood, overlooked in the interests of innovation

(and economy). I have included Dawes's post-Grenfell analysis within this final part of the report because she introduces a distinction between the "setting of standards", and the activity of "regulating". Her view suggests that her department did not regard itself as a regulator, but nevertheless felt that it could set standards. **(31)**

Despite the excellence of Bisby's 300-page analysis, the UK Government commissioned another study, described as an independent review of the Construction Products Testing regime. The Grenfell Inquiry uncovered many shocking examples of the failures of the test regime or how it was "gamed" by manufacturers to allow unsuitable and unsafe products to be approved. Morrell, a past construction advisor to the Government and Day, a lawyer, went into considerable detail about the nature of building standards, regulation and testing, though avoiding a critical analysis of current testing by bodies such as the British Board of Agreement. However, they do comment on the ineffectiveness of the regulations but their analysis seems to be very sympathetic to the risks taken by manufacturers

> …the evidence of the Grenfell Tower Inquiry demonstrates the Regulations' ineffectiveness in serving the larger purpose of building safety. Whether that was the result of expecting the Regulations to serve a purpose they were never designed for, or of faults in the system, or of matters of institutional or individual default… .
>
> The fire at Grenfell Tower illustrates just how much knowledge about the behaviour of buildings in fire needs to be accumulated and disseminated, and how much that knowledge needs to be a public good.
>
> …the impact of any new regulatory requirements relating to construction products, including the general safety requirement, on the availability and cost of product liability insurance needs to be studied in advance of implementation. The importance of this is two-fold: whilst the intention may be for manufacturers to minimise their risk by doing what is required in the regulations, rather than insuring themselves against it, without such cover redress for anyone harmed is backed only by the balance sheet of the manufacturer; and prudent manufacturers will think long and hard about the balance of risk and reward for products where the risks may be high, and the returns low. **(32)**

Despite the importance of the chemical emissions and health effects of building products, and the ongoing debate in the EU about changes to the Construction Product regulations (CPR) Morrell and Day appear to have been unaware of this important aspect of the safety of building materials, or it was not included in their brief from Government. The importance of chemical emissions was ignored in their analysis of testing.

> For products covered by the CPR, the moment for that truthfulness is at the Declaration of Performance prior to a product being placed on the market. …where thought is also given as to whether all products might be required to make such a declaration.

This important report fails to address the issue of privatisation, though deregulation gets a brief mention, but it does provide a clear indication of the inadequate systems in place in the UK compared with other countries. They do mention New Zealand (BRANZ) which commissions research and the Deutsches Institut für Bautechnik (DIBt) in Germany as examples of where there are technical authorities and centres of excellence which are sadly lacking in the UK. These roles had previously been performed by the Building Research Establishment (BRE) which was privatised. BRANZ, for instance carries out significant work on indoor air quality and DIBt publishes reports on subjects such as radiation from construction materials (in English!) **(33)**.

Energy efficiency ignores indoor air quality

Given the performance gap failings of modern building to achieve better insulation standards, many architects and public bodies have turned to more extreme energy standards such as "Passivhaus" **(34)**. Devotees of passivhaus exhibit an almost religious devotion to this system which was set up in Germany in the early 1990s. Passivhaus is officially "materials neutral" but was quick to embrace chemical methods of building, promoting the concept of claimed ultra-low insulation and wrapping buildings tightly with plastic membranes. Passivhaus buildings have to introduce mechanical ventilation systems to cope with the extreme airtightness as there is a lack of natural ventilation. Many German architects were hostile to passivhaus from the start and founded a more holistic and natural approach called Aktivhaus **(35)**.

The Active House has a much greater adherence internationally, **(36)** but passivhaus has remained popular in the UK and Ireland where efforts have been made for it to become a formal requirement for all buildings despite the fact that it is a private commercial business. Proposals in Scotland to adopt the passivhaus approach have recently been incorporated into energy standards. **(37)** Passivhaus claims to achieve good indoor air quality through the use of mechanical ventilation but does not appear to have policies about reducing the use of chemicals.

Some radical green and environmental campaigners in the UK seem confused about chemical and plastic materials. Activist groups such as "Insulate Britain" and "Just Stop Oil," who sometimes glue themselves to roads and other physical structures, use petro-chemical based glues such as Cyanoacrylate, or C5H5NO2, Cyanoacetate, Formaldehyde, and Methanol, which may lead to health problems in the future. (with apologies for quoting the Daily Mail on this!) **(38)**

Chemicals are with us in all aspects of society, and we are subject to greater exposure to their emissions as we spend much of our time indoors. Some awareness of these problems happened as a result of the Covid pandemic, with people spending even more time trapped in their homes. This has led to new research into the way in which some materials not only trap mould but other pathogens related to Covid and other diseases. Unfortunately, this has led to even more attempts to incorporate hazardous chemicals such as anti-microbials and fungicides into building products and other chemical cleaning products.

There is growing pressure for the renovation of existing houses and campaigns for higher levels of insulation in an attempt to reduce heating bills and to reduce carbon emissions. This could lead to higher concentrations of hazardous chemicals inside houses and other kinds of buildings such as schools, that can have a major impact on the health and development of occupants, particularly children. It is frequently stated that most people spend at least 90% of their time inside, and so this does mean that there has been some increased interest in indoor air quality and emissions from hazardous materials.

There is little doubt that the built environment has a significant effect on human health. This is caused by both external and internal pollution. There are a wide range of sources including traffic and industrial pollution, agricultural sources such as pesticides, chemicals in foods and packaging, cleaning materials and a range of air fresheners and cosmetics as well as building and interior finishes and furnishing. Given the complexity of this "toxic cocktail" problem and the difficulty in understanding it, the response of many professionals and scientists is to ignore it. If they cannot find an easy answer then they focus on those where the problem seems simpler. For instance, in the UK the focus on pollution has been almost entirely on vehicle emissions. There is an assumption that there is good data available on traffic pollution and also good evidence of the negative health effects. As a result, 95% of research funding on air quality has focussed on traffic pollution, to the extent that the same data and research is produced over and over again. When speaking at conferences on air quality the author has found himself as the only speaker on indoor air quality while the other 7 or 8 are talking about traffic pollution.

For reasons that are still not clear, the vast majority of academics, scientists and policy makers in the UK have decided that indoor air quality is mainly a problem of cooking, personal hygiene, cleaning products and emissions from wood burning stoves and this is critically examined in Chapter 4. Indoor pollution from building materials remains side-lined, even in more recent research programmes under the Clean Air banner. In this book, emissions from building materials including Volatile Organic Compounds (VOCs), formaldehyde particulates, flame retardants, PFAS (per- and polyfluorinated substances), isocyanates and many more are examined. Not only are the problems of indoor emissions from building materials inadequately examined but official bodies have deliberately decided to give them minimal consideration. Evidence of building materials emissions is widely available in international research but this is largely ignored in UK research.

The underlying principle behind this book is the idea that a holistic approach should be adopted when designing and considering sustainability and health impacts in building. The standard approach to building and design unfortunately is to cherry pick aspects of green building and largely ignore this holistic approach. While energy efficiency continues to be given the bulk of attention, health in buildings is left off the agenda. For a time, there was a lot of discussion of sick building syndrome (SBS), a term familiar to most people, but it is unusual to hear this mentioned today. The issue of SBS was largely seen as a question of ventilation and air conditioning. NHS guidance suggests that SBS is still poorly understood but particularly by the writers of the NHS guidance. This can be compared with much more sophisticated advice in the USA **(39) (40)**.

One of the main contributors to poor indoor air quality (apart from poor ventilation) is the way in which the building is constructed and the materials used to construct it. Thus, this book largely focuses on what is known about construction materials and how these might create health problems. Rather than stating that certain materials are good or bad this judgement is left to the reader to base their choice and specification decisions on the best information that it has been possible to present here.

If in doubt, the best approach to adopt is that of the "PRECAUTIONARY PRINCIPLE." If there appear to be serious concerns, even if reliable scientific information still leaves some doubt, then surely it is better to avoid the risk and go for an alternative that is substantially less risky. Fortunately, today, there are plenty of building methods and materials available that are very low risk, not necessarily the situation when the Green Building Digest was written in the 1990s **(41)**.

Some of the book deals with wider issues than those specifically related to building materials. Mould and damp for instance is an issue related to building materials but is also a complex cause of health problems. It has also been important to analyse Government and professional policies on these issues in an effort to understand why health and indoor air quality in buildings has been so determinedly sidelined. There is a general acceptance of chemicals in building materials without realising the dangers of their effects on health.

If there are pollutants in our homes that can harm our health, the first thing we should do is try to reduce them appearing in the first place, a process known as **source control,** and also to improve ventilation. Conventional thinking within the construction industry is to address ventilation first and ignore source control. Chapters 2 and 3 discuss a range of materials normally used in buildings, how they contain chemicals and what effect this might have on health. Chapter 6 addresses ventilation.

We do not know enough about the long-term effects of many of the chemicals with which we surround ourselves. What can be said unequivocally is that some of these chemicals are known to be dangerous, and many more are likely to be hazardous to an extent. We do not know how they affect people when mixing and reacting with other chemicals, how they affect people over the long term or how they affect more vulnerable people, and so it makes sense to adopt the precautionary principle and look carefully at reducing the overall scale of this insidious chemical load on occupants until such time as we are sufficiently clear that there is no risk.

Volatile Organic Compounds are a loose group of thousands of different compounds all of which contain carbon (hence "organic") and have relatively low boiling points (hence "volatile") and are therefore readily emitted as a gas in most internal conditions. Some are safe, some are known carcinogens and many appear to have long-term effects on our health. They are usually grouped into three sub-types relating to boiling point, being "Very volatile" (VVOC), (Normal) VOC and "Semi-volatile" (SVOC). When monitoring properties, many sensors take an aggregate value for a range of VOCs and the phrase "Total VOC" (TVOC) is often seen as a catch-all figure for VOCs generally.

We need more houses?

Politicians are always launching programmes to build more houses to deal with the homeless problem, especially as elections grow nearer. The reality is that these programmes are rarely achieved, and most homeless people cannot afford them anyway. Arguments rage about whether to relax planning restrictions, whether to build on green belt, concrete over the countryside or brownfield sites and so on **(42)**. Some politicians are even calling for healthier homes, and this has crept up the agenda as discussed in Chapter 4.

Government Ministers constantly clutch at any new ideas as to how housing production and quality can be improved, particularly if it can also cut costs. This led to the disaster of high rise and concrete system buildings in the 60s, many of which had to be demolished. Those that remained like Grenfell Tower became part of the **retrofit disaster** problem. In recent years, a new euphemism for system building has emerged called Modern Methods of Construction (MMC). MMC means many different things but all suggest that buildings can be prefabricated off site, using timber systems and even concrete panels, which are making a comeback.

In fact, MMC is another potential disaster (discussed in Chapter 8) and many MMC companies have gone out of business with brand new buildings failing, such as three schools in Somerset, Northampton and Harlow, even before they were occupied **(43)**.

Dominating the news at the start of the school term in September 2023 has been the realisation that hundreds of school buildings, hospitals and other public buildings, like court houses, have had to be shut down due to the danger of collapsing RAAC construction. Reinforced Autoclaved Aerated (Foamed) Concrete (RAAC) was introduced mainly in public buildings in the 1950s but concerns about its structural performance were raised as early as 1982 and production of RAAC was stopped then **(44) (45)**.

As Chloe McCulloch points out, in *Building Magazine*, the current scandal about hundreds of schools being closed due to the risk of collapsing roofs building with RAAC is an example of a deeper malaise in Government policy about buildings and building materials.

> The government has known for five years that reinforced autoclaved aerated concrete, or RAAC for short, posed a risk. In 2018, a school ceiling in Kent made of RAAC panels

collapsed, and it was only sheer good luck that no one was hurt as it happened during a weekend when the school was not open. **(45) (46)**

While RAAC concrete may not have contributed to poor indoor air quality it has certainly put the lives of many school children and others at risk. But the reason for highlighting MMC and RAAC here is as examples of builders and investors clutching at innovative forms of construction that were not tried and tested. This indicates a fundamental culture about building that puts people and their health and safety last and so it is hardly surprising that indoor air quality and emissions from building materials is low on the agenda or even deliberately ignored by scientists who try to distract attention by discussion of cooking, cleaning or wood burning. The aim of this book is to challenge this and to establish indoor air quality and building materials emissions as a key issue in modern society.

References

(1) The benefits of chemicals in everyday life. (n.d). Available at: www.shell.com/business-customers/chemicals/the-benefits-of-chemicals-in-everyday-life.html [Accessed 25 Sep. 23].
(2) Global chemicals outlook. (2013). Available at: www.unep.org/explore-topics/chemicals-waste/what-we-do/policy-and-governance/global-chemicals-outlook [Accessed 25 Sep. 23].
(3) Chemscore report. (2022). Available at: https://chemscore.chemsec.org/ [Accessed 25 Sep. 23].
(4) Lakhani, N. (2023). Be vigilant, hold your ground: Erin Brokovitch rallies Ohio town after train disaster. Available at: www.theguardian.com/us-news/2023/feb/25/be-vigilant-hold-your-ground-erin-brockovich-rallies-ohio-town-after-train-disaster [Accessed 25 Sep. 23].
(5) Mark Ruffalo on playing the lawyer who took on DuPont. (2019). Available at: https://calgary.citynews.ca/2019/11/20/qa-mark-ruffalo-on-playing-the-lawyer-who-took-on-dupont/ [Accessed 25 Sep. 23].
(6) Water companies could face legal action. (18.11.2021). Available at: www.ofwat.gov.uk/joint-ofwat-environment-agency-and-defra-announcement-november-2021/ [Accessed 25 Sep. 23].
(7) Meet Arlene Blum PhD (n.d). Available at: www.arleneblum.com/ [Accessed 25 Sep. 23].
(8) Kramer, K. (3.11.2021). In situ with Arlene Blum. Available at: www.chemistryworld.com/culture/arlene-blum-our-summit-is-reduced-toxics/4014555.article/ [Accessed 25 Sep. 23].
(9) Ashrae: In memory of Hal Levin. (n.d). Available at: www.ashrae.org/news/esociety/in-memory-of-hal-levin [Accessed 25 Sep. 23].
(10) Charles, J. Weschler PhD Ruitgers Environmental and Occupational Health Sciences Institute (n.d). Available at: https://eohsi.rutgers.edu/eohsi-directory/name/charles-weschler/ [Accessed 25 Sep. 23].
(11) Adam, D. Earthshakers: The top 100 green campaigners of all time. (2006). Available at: www.theguardian.com/environment/2006/nov/28/climatechange.climatechangeenvironment [Accessed 25 Sep. 23].
(12) Douthwaite, R. ed. (2003) Before the Wells Run Dry. Feasta.
(13) FEASTA (n.d). Available at: https://www.feasta.org/ [Accessed 25 Sep. 23].
(14) Cancer Prevention and Education Society (n.d). Available at: https://www.cancerpreventionsociety.org/ [Accessed 25 Sep. 23].
(15) UCLan professor honoured by Fire Brigades Union for life-saving research (2023). Available at: https://www.uclan.ac.uk/news/fbu-solidarity-medal [Accessed 25 Sep. 23].
(16) Dr. Michelle Bellingham (n.d). Available at: www.futurebuild.co.uk/guest/dr-michelle-bellingham-glasgow-university-author-of-the-royal-college-of-obstetricians-report-on-chemical-exposures-during-pregnancy/ [Accessed 25 Sep. 23].
(17) Booth, R. (2021). Grenfell resident who raised fire concerns labelled troublemaker, inquiry told. www.theguardian.com/uk-news/2021/apr/21/grenfell-resident-who-raised-fire- [Accessed 25 Sep. 23].

(18) Holgate, S. et al. (2020). The inside story: Health effects of indoor air quality on children and young people. RCPCH. www.rcpch.ac.uk/sites/default/files/2020-01/the-inside-story-report_january-2020.pdf [Accessed 25 Sep. 23].
(19) Collin. J. et al. (2021). Conflicted and confused? Health harming industries and research funding in leading UK universities. BMJ. Jul. www.bmj.com/content/374/bmj.n1657 [Accessed 25 Sep. 23].
(20) Energy House 2.0 Project. A world leading energy performance test facility (n.d). https://energyhouse2.salford.ac.uk/ [Accessed 25 Sep. 23].
(21) Woolley, T. et al Green Building Handbook Volumes 1 and 2. Taylor and Francis.
(22) Havre, J. (2022). ANSI blog: Metrics of sustainability: Embodied carbon vs. energy efficiency (2022) https://blog.ansi.org/?p=170052 [Accessed 25 Sep. 23].
(23) UKGBC Published the Net Zero Carbon Buildings Framework in 2019. https://ukgbc.org/wp-content/uploads/2019/04/Net-Zero-Carbon-Buildings-A-framework-definition.pdf [Accessed 25 Sep. 23].
(24) Grenfell Tower Pubic Inquiry. Witness Statement of Philip John Heath. https://assets.grenfelltowerinquiry.org.uk/KIN00020709_2019.10.15%20Witness%20Statement%20of%20Philip%20John%20Heath.pdf [Accessed 25 Sep. 23].
(25) Chinchalkar, S. (2021) What is Celotex made from? https://insulation4less.co.uk/blogs/insulation-guides/what-is-celotex-made-from [Accessed 25 Sep. 23].
(26) Usable Buildings for Feedback and Strategy (25 Sep. 2023). https://www.usablebuildings.co.uk/ [Accessed 25 Sep. 23].
(27) Zero Carbon Hub closes following green policy changes (2016). https://www.edie.net/zero-carbon-hub-closes-following-green-building-policy-changes/ [Accessed 25 Sep. 23].
(28) Performance Gap (n.d). www.thebuildingshub.co.uk/resources/performance-gap/ [Accessed 25 Sep. 23].
(29) Gorse, C. et al. (2015). Addressing the thermal performance gap: Possible performance control tools for the construction manager. Proceedings 31st Annual ARCOM Conference, 31, 337–346.
(30) Apps, P (2022). Show me the bodies: How we let Grenfell happen. One World 2022.
(31) Bisby, L. (2021). Phase 2 – regulatory testing and the path to Grenfell. Grenfell Tower Inquiry. The University of Edinburgh School of Engineering 10th November 2021. https://assets.grenfelltowerinquiry.org.uk/LBYP20000001_Professor%20Luke%20Bisby%20Phase%202%20Report%20 %20Regulatory%20Testing%20and%20the%20Path%20to%20Grenfell_1.pdf [Accessed 25 Sep. 23].
(32) Morrell, P. and Day, A. (2023). An independent review of the Construction Products Testing regime. Department for Levelling up Housing and Communities.
(33) Morrell (2023). Op cit.
(34) What is Passivhaus? (n.d). www.passivhaustrust.org.uk/what_is_passivhaus.php [Accessed 25 Sep. 23].
(35) Hegger M. et al. (2016). Aktiv Haus: From Passivhaus to Energy-plus House. Birkhauser.
(36) Active House National Alliances. (n.d). www.activehouse.info/ [Accessed 25 Sep. 23].
(37) Riding, J. (2023). Passivhaus now: Scotland's plan for tough new building standards. www.insidehousing.co.uk/insight/passivhaus-now-scotlands-plan-for-tough-new-building-standards-81499#:~:text=From%20January%202025%2C%20all%20new%20build%20homes%20will%20have%20to,cheap%20for%20residents%20to%20heat [Accessed 25 Sep. 23].
(38) Just stop oil using adhesives made from fossil fuels. (2022). www.dailymail.co.uk/news/article-11465261/How-hypocrite-Just-Stop-Oil-activists-using-adhesives-FOSSIL-FUELS.html [Accessed 25 Sep. 23].
(39) NHS. Sick Building Syndrome (n.d). www.nhs.uk/Conditions/Sick-building-syndrome/Pages/Introduction.aspx [Accessed 25 Sep. 23].
(40) US EPA Indoor Air Quality. www.epa.gov/iaq/pdfs/sick_building_factsheet.pdf [Accessed 25 Sep. 23].
(41) How bio-based materials are revolutionising sustainable construction (n.d). www.tunley-environmental.com/en/insights/how-bio-based-materials-are-revolutionising-sustainable-construction#:~:text=Some%20products%20which%20are%20currently,walls%2C%20floors%2C%20and%20roofs [Accessed 25 Sep. 23].

(42) Sunak confirms bulk of Britain's new homes will not concrete over the countryside (2023). www.independent.co.uk/news/uk/home-news/rishi-sunak-housing-plan-uk-michael-gove-b2380605.html [Accessed 25 Sep. 23].
(43) Third Caledonian Modular School ordered to close over structural safety concerns. *PBC Today* (2023) www.pbctoday.co.uk/news/health-safety-news/third-caledonian-modular-school-ordered-to-close-over-structural-safety-concerns/131692/#:~:text=The%20closure%20of%20the%20Northampton,into%20administration%20in%20March%202022 [Accessed 25 Sep. 23].
(44) Ridings, J. (2023). How to locate RAAC and what to do about it. *RICS Journal*. ww3.rics.org/uk/en/journals/built-environment-journal/locating-raac-reinforced-autoclaved-aerated-concrete.html [Accessed 25 Sep. 23].
(45) McCulloch, C. (2023). Scandal over RAAC closures points to a deeper malaise. www.building.co.uk/comment/scandal-over-raac-school-closures-points-to-a-deeper-malaise/5124933.article#:~:text=The%20government%20has%20known%20for,the%20school%20was%20not%20open [Accessed 25 Sep. 23].
(46) Zhou, G. (2023). A review on durability of foam concrete. *Buildings*, *13*(7), 1880. www.mdpi.com/2075-5309/13/7/1880 [Accessed 18 Sep. 23].

2 Chemicals in building materials

So extensive is the use of chemicals in building materials today, it is only possible to provide a summary in this book. Chemicals listed here, by and large, are hazardous and can find their way into the indoor environment of most buildings. The list at the end of this chapter (Table 2) provides a summary, but it is not always easy to find this information as manufacturers like to plead commercial confidentiality when asked what is used in their products.

Where hazardous chemicals are used, they should be registered on the REACH directive. Even though the UK is no longer a member of the EU, a great deal of EU legislation still applies and there is a draft bill before Parliament called the "Consumer Protection, Environmental Protection Health and Safety, The REACH (Amendment) Regulations 2023." How this will be applied and administered is not clear but companies producing building materials in the UK do continue to register the chemicals they use with REACH. **(1)**, **(2)**

Accessing information about chemicals, it is necessary to know the REACH registration number and not surprisingly chemical companies and product manufacturers do not make this easily available. You may find reference to REACH registration on product data sheets and also a statement that products are not hazardous and do not need to be registered with REACH. However, the classification of mineral fibres, for instance, is controversial and is discussed in more detail below as not everyone would accept the assurances provided by Rockwool.

> UK health and safety regulations (including REACH) do not require a Safety Data Sheet (SDS) to be provided for mineral wool insulation. However, MIMA, the Mineral Wool Insulation Manufacturers Association, members voluntarily make REACH compliant safety data sheets available for their products to ensure that health and safety information is provided in a recognised standard format.
>
> Owing to their high bio-solubility, the fibre types of ROCKWOOL stone wool insulation materials are assessed as free from suspicion of possible carcinogenic effects in accordance with EU Directive 97/69/EC (Note Q). In October 2001, the International Agency for Research on Cancer (IARC) classified mineral wool insulation as Group 3 (not classifiable as to its carcinogenicity in humans). i.e. not classified as possibly carcinogenic to humans. **(3)**

While REACH might seem to be a powerful tool to protect people from the effects of hazardous chemicals it has been argued that industry lobbying tried to limit the restriction of chemical products. Industry lobbying during the legislative process of REACH between 2001 and 2007

DOI: 10.1201/9781003129226-2

was extensive. During these years, the chemical industry orchestrated what Wenger et al called the largest ever industry lobbying campaign in Europe.

> The purpose of the new regulation REACH was to protect human health and the environment from the risks that can be posed by chemicals through the promotion of alternative test methods. REACH shifted the responsibility of assessing and managing the risks posed by chemicals from the authorities to the industry,
>
> Based on REACH, chemicals exceeding one metric ton, produced or imported by a company annually, must be registered with the European Chemical Agency (ECHA), following a technical dossier and a chemical safety report. **(4)**

Wenger et al tell a fascinating story of the "Battle of REACH" in which corporate lobbying from the third biggest industrial sector in Europe managed to limit the powers of the EU to control the chemicals industry:

> REACH is an example of politicization of business where the Commission tried to enforce regulation on a relatively untethered market and the chemical industry, with its capable influence as a policy shaper within public and private institutions, put pressure on EU officials, ministers and other government officials in order to counteract support for by promoting REACH as a threat to economic prosperity as opposed to a threat to public health. The example of REACH illustrates a blurring line between state and corporate lobbying…but one thing is clear, the opponents of REACH have managed to significantly reduce the regulatory demands that would otherwise have been imposed upon the industry.

Health and safety data sheets can give the impression that chemicals or materials used are not hazardous, as though this is based on independent assessment, whereas it may be the manufacturer themselves that has come to such conclusions.

Looking for simple guidance

When debating the issues of building material emissions with organisations providing guidance on healthy homes and buildings, some have stated that architects and specifiers do not want to wade through detailed lists of chemicals. They just want a simple classification system, even though this does not exist. Standard specification tools used by architects and even "green" guides to building materials do not provide much detail in terms of chemical constituents or potential hazardous to health. "Just tell us what to avoid" was one statement even though that could mean avoiding the majority of products! This book and this chapter will not be easy reading for those looking for simplistic guidance.

Common materials that contain chemicals

The most accessible way to deal with the issue of chemicals in building materials is to identify a number of common elements and products. These include gypsum plasters and plasterboard (drywall), glues, sealants and fillers, composite boards, plastics, floor screeds and floor finishes, paints, air tightness membranes and tapes, insulations, foam and fibre and even natural products that include chemicals. Also, basic construction materials such as concrete blocks are no longer just made with cement and sand and stone but also include chemical additives.

Floor finishes and carpets

Vinyl flooring is popular as it is cheap. Available in sheet and tile form it is widely used in homes but also public buildings such as hospitals. Vinyl is made from PVC and dioxins are emitted during the manufacture of PVC, which are damaging to the wider environment. Vinyl also contains Ethylene Dichloride and vinyl chloride and pthalates which are also hazardous chemicals. Pthalates are dangerous and attempts have been made to ban phthalates from use in children's toys and products that come into contact with food and drinks. Vinyl flooring can also emit VOCs and there can be serious emissions from adhesives used to stick the flooring down. Care should be taken when removing old vinyl tiles as asbestos was used as a backing material in the past. **(5)**

> Exposure to Volatile Organic compound (VOC) concentrations greater than 3.0 mg/m3 leads to irritation, headache, etc. Significant concentrations of VOC are achieved mainly by vinyl flooring and nylon carpet. VOC Concentrations of other floor coverings are several times smaller. In the case of vinyl flooring, after the initial steep growth (3.30 mg/m3), Total VOC decreases by up to one third in the monitored time (0.86 mg/m3). TVOC concentrations also rise steeply in the case of vinyl flooring. **(6)**

Timber flooring is seen as a healthier alternative, but VOCs can be emitted from composite timber boards and panels. This can be due to the glues and adhesives used in the product and when fixed to the subfloor. Highly toxic chemicals such as naphthalene have been found in flooring adhesives. **(7)**

> In the case of hardwood flooring, acetic acid, acetone, decanal, octanal, pentadecane, tetradecane and toluene are monitored. Of these compounds, significant concentrations only reach acetic acid. The maximum concentration of acetonic acid is approximately half that of linoleum. However, the decrease in time is slower. The acetic acid trend follows the concentration of the TVOC. Concentrations of other substances of concern are very low. M. Kraus 2020 op cit **(6)**

Carpets can present a wide range of indoor air quality problems. Often the main problem is caused by backing materials such as styrene-butadiene, rubber and latex as well as adhesives used to stick the carpet or tiles down. Some carpet tiles are still made with a bitumen backing.

VOC emissions problems are often perceived as occurring shortly after the carpet has been installed and the "new" smell gradually dissipates. However, this does not mean that the hazardous chemicals have gone. Emissions can include Styrene and 4-phenylcyclohexene (4-PCH).

> …carpet with a polyvinyl chloride backing emitted formaldehyde, vinyl acetate, isooctane, 1,2-propanediol, and 2-ethyl-1-hexanol. Of these, vinyl acetate and propanediol had the highest concentrations and emission rates. The carpet with a polyurethane backing primarily emitted butylated hydroxytoluene. With the exception of formaldehyde, little is known about the health effects of these VOCs at low concentrations. **(8)**

Carpets are made from both synthetic plastic materials and from natural wool and many will contain flame retardants. The fact that small children crawl round floors a lot and thus have a greater chance of being affected by the chemicals in the flooring materials is of concern. In a project carried out by the Women's Environmental Network, chemicals were found in the majority of samples taken.

> As part of the investigation, researchers from the Universities of Amsterdam and Notre Dame and the Ecology Centre tested samples of products sold by the seven largest carpet companies in Europe – Interface, Desso (Tarkett), Beaulieu International Group, Balta Industries, Milliken, Associated Weavers and Forbo – for a number of different chemicals. In total, 12 of the 15 samples were found to contain chemicals of concern. **(9)**

Carpets can contain chemical stain repellent treatments, PFAS and anti-microbials including Triclosan and Formaldehyde, Isocyanates, Phthalates, fly ash, flame retardants as well as a range of adhesives including endocrine-disrupting chemicals according to The Healthy Building Network (HBN) who have identified "44 toxic chemicals commonly used in carpets and carpet adhesives." **(10)** Ceramic tile and slate flooring may seem to be emission free, but there may still be emissions from the adhesives and cements used to stick them to subfloors.

Paints and varnishes

Decorating finishes in buildings are a major source of chemical emissions and can produce significant odours when first applied. Building emissions deniers like to point the finger at paints as these are seen as the only and short-term source of VOCs, which will soon dissipate. Also, paints have been subject to greater control, which has led to manufacturers shifting to water-based substances, reducing the use of more dangerous solvents. However, this does not mean that water-based paints are therefore safe and free of chemical emissions. The majority of paints are made of synthetic plastic chemicals, and while they might not smell as bad they still present a hazard. Paints are often said to emit Semi Volatile Organic compounds (SVOCs).

> It was found that all paints contain SVOCs in order to circumvent solvent classifications. Even though the spectrum of emitted substances was limited under test chamber conditions, the emission potential was very high due to the increased release of individual VOCs/SVOCs. Additionally, the paints released heightened acetic acid levels and are therefore not favoured for the use in sensitive environments. **(11)**

Chemicals are still added to paints to accelerate drying, to create greater adhesion and to have a fungicidal affect to reduce the risk of mould growth. Other chemicals are added to make the paints easier to clean and some manufacturers even claim that paints can be air purifying, absorbing VOCs and chemicals from the indoor air. If these additives are not classified as VOCs, there are few restrictions.

Some people have allergic reactions to emulsion paints.

> Inhalation of paint fumes can lead to a runny nose, sore throat, cough and nasal congestion as common symptoms along with irritated and watery eyes. On the other hand, people who experience reactions from direct contact can have a localized skin irritation, rash or discoloured skin, itchiness and sometimes blistering. Burning and swelling are also common symptoms of paint allergies via direct contact. **(12)**

> It is possible to get allergic reactions due to the substances that are present in acrylic paint fumes. Some acrylics are known to include a small amount of formaldehyde or other

substances that delay mould growth. Formaldehyde and similar mediums are responsible for some people developing allergies. **(13)**

Eggshell paints are acrylic and will contain Titanium Dioxide, 1,2-benzisothiazol-3(2H)-one and C(M)IT/MIT (3.1) and Methylisothiazolinone according to a typical data sheet. **(14)** Gloss paints can contain alkyd polymer resins made from a polyol such as propane 1,2,3-triol(glycerol) with a diabasic acid such as benzene-1,2-dicarboxylic (pthalic) anhydride and a drying oil made from linseed or soya. **(15)** Seventy percent of the content of most paints is Titanium Dioxide which is even used in so called green or eco paints. According to the New Jersey Department of Health Hazardous Substance Fact Sheet, Titanium Dioxide can have acute health effects and even be carcinogenic.

> Exposure can irritate the eyes, nose and throat. The following chronic (long-term) health effects can occur at some time after exposure to Titanium Dioxide and can last for months or years: Cancer Hazard: Titanium Dioxide may be a carcinogen in humans since it has been shown to cause lung cancer in animals. **(16)**

It's a mistake to assume that just because paints and finishes are water-based that they are chemical and emissions free.

Synthetic, mineral fibre insulations

Glass fibre insulation and other mineral wool insulations are possibly the most widely used insulation materials in buildings. The majority of lofts in houses will have pink or yellow or grey and dirty glass fibre. Most landlords and energy efficiency schemes will recommend installing extra glass fibre.

Anyone who has handled glass fibre will be aware that it is not very pleasant, leaving skin itchy and some people can have allergic reactions. A number of efforts were made to ban glass fibre in the 1970s and some years later as it was considered as carcinogenic. After a big campaign by the industry the various public health bodies were forced to withdraw their health concerns and so glass fibre insulation is still with us today. The Clinton administration in the USA tried to ban glass fibre but had to back down in 1993. **(17)**

A debate took place in the pages of the *American Journal of Industrial Medicine* with extensive references to the epidemiology of fibrous glass and lung cancer, with scientists repeating the claim that glass fibre is carcinogenic.

> In response to our commentary on fibrous glass and cancer [Infante et al., 1994], three letters have been received by the journal. The arguments put forth in these letters do not lead us to alter our scientific view that fibrous glass insulation is carcinogenic. **(18)**

Using the precautionary principle specifiers, householders and builders may wish to avoid this material and indeed it is not unusual to find building workers who are unwilling to handle it. Despite this it is common to find green campaign organisations, promoting energy efficiency, showing photos of glass fibre being installed in lofts.

A 1997 European Directive (97/69/EC) classified mineral wool fibres as hazardous substances. This classification was based on their chemistry, size and bio-persistence. According to this European Directive, the refractory ceramic fibres are classified as category 2 (substances that should be regarded as if they are carcinogenic to man) and are labelled with risk phrases

R49 (may cause cancer by inhalation) and R38 (skin irritant) while the mineral wools are classified as category 3 (substances which cause concern for man owing to possible carcinogenic effects. **(19)**

The controversy about the health dangers of glass fibre will continue with the Man Made Mineral Fibre (MMMF) lobby continuing to fight against restrictions. Meanwhile a leading mineral fibre manufacturer, Knauf, changed the formulation of one of its products which it branded as "Earthwool." It was not uncommon for some green campaigners to claim that "Earthwool" was a natural material made from earth with natural and recycled fibres using a technology known as ECOSE but Knauf later withdrew the name.

> As part of our continuous efforts to improve our insulation product offering, we have decided to **discontinue Earthwool** from all of our branding. This change **does not** impact the performance of any of our products. Based on your (sic) feedback, it was clear that the use of Earthwool, alongside Knauf Insulation, was causing some confusion; whether that be in specifications, product choices, or in general when referring to Knauf Insulation. With this in mind, as of 1st April 2021, we will discontinue Earthwool. **(20)**

Various versions of glass-based materials are still made from molten glass (possibly waste glass) with an added binder. Some binder bonding agents are said to be derived from plant material but health and safety data sheets still warn that the material can cause itching. Bonding agents may still involve hazardous chemicals.

Stonewool

The other main form of mineral fibre insulations is made from melted rock, rather than glass. These can be referred to by the generic name Stonewool but products are often referred to as Rockwool, one of the main producers. These products can come in a wide range of forms with different densities. They are claimed to provide good thermal and acoustic properties and also to be fire safe. Stone wool materials are sometimes used as fire stopping in composite constructions as well as insulation quilts and batts.

Stone wool involves melting stone at 1600 degrees, which is then spun with a synthetic binder and other materials. Some products are said to be made from recycled materials from blast furnaces. There is much debate about whether stone wool products contain formaldehyde. Emissions from a new stone wool factory in the USA declared potential formaldehyde emissions from the factory flue. This new factory, being constructed to produce stone wool products in West Virginia, had to make a declaration to achieve various air and environmental permits in the face of significant protests from the local community against the factory. **(21)**, **(22)**

The list above is of gas emitted, so tons of gas is a significant amount.

Table 2.1 Estimated potential discharges of regulated air pollutants

Nitrogen Oxides: 239 tons per year
Sulphur Dioxide: 148 tons per year
Carbon Monoxide:74.1 tons per year
Volatile Organic Compounds: 472 tons per year
Formaldehyde: 67.6 tons per year
And many other chemicals

Some campaigning websites claim that Stonewool is harmful to health and a number of scientific studies have identified levels of formaldehyde and VOCs, though it is argued that emissions in buildings are below international safety limits. **(23)**

More recent studies discuss low formaldehyde with Rockwool stating that their products do not use "added" phenol formaldehyde. It is possible that there will still be formaldehyde which is naturally occurring from the manufacturing process. **(24)**

> Standard ROCKWOOL products do currently use a phenol formaldehyde binder during production. During manufacturing, the binder is cured at very high temperatures leaving only trace amounts in the product after it is produced. ROCKWOOL AFB®, SAFE'n'SOUND®, and COMFORTBATT® have all achieved GREENGUARD GOLD status for indoor air quality. ROCKWOOL **also offers** a no added formaldehyde product, AFB evo™, starting July 1st 2017. This product is produced with a new formaldehyde free binder, and retains all of the properties of the original AFB® product. **(25)**

Leaving aside the question of formaldehyde there remain concerns about the dangers of the fibres released from stone wool and other mineral fibre/wool products. It was claimed that the fibres are not dangerous in a Bridgend council planning report, related to the expansion of a stone wool factory in South Wales, though the source of this was a body known as the EUCEB.

> Is claimed to demonstrate that the fibre is sufficiently bio-soluble to pose no risk to human health if inhaled. Any fibre accidentally inhaled will be readily dissolved and cleared by the body's natural defences. **(26)**

The EUCEB (certification board for mineral wool products) giving these assurances claims to be a not for profit "independent body." However, it lists its members as Saint Gobain, Knauf, Owens Corning, Fiberglass, Rockwool, Roxul and Steinull (Iceland). The management board of EUCEB consists of employees from Saint Gobain, Knauf and Rockwool. They claim to use an independent sampling organisation to carry out tests, but it has not been possible to find out who carries out this work. **(27)**

Medical investigations into mineral fibres suggests that they can be dangerous and toxic. There are different mechanisms in which the toxicity receives different human response depending on the three carcinogenic mineral fibres: crocidolite, chrysotile and erionite. **(28)**

> Toxic elements, such as Chromium, Barium, and Nickel are present in mineral wools, but in low concentrations (<0.2%) as well as organic resin, which may decompose into ammonia upon heating. **(29)**

Mineral fibre insulations are said to be non-combustible and so do not require the addition of flame-retardant chemicals. Stone wool is said to resist temperatures of over 1000 degrees and does not melt until 2000 degrees are reached but independent verification of this could not be found. Following the Grenfell disaster many specifiers switched to the use of stone wool.

Polystyrene

Polystyrene foam contains the chemical styrene, which has been linked to cancer, vision and hearing loss, impaired memory and concentration and nervous system effects. **(30)**

Styrene is relatively stable once formed into insulation but it can emit Pentane, a hazardous blowing agent and flame retardants. The flame retardants are rarely divulged openly described, for instance, as a "polymerised flame retardant". **(31)**

Brominated flame retardants (Hexabromocylododecane) have also been used in polystyrene.

The Pentane blowing agent can be inhaled and absorbed by the skin. It can irritate and burn skin and eyes, nose, throat and lungs. Exposure can cause headache, dizziness and damage the nervous system. It is flammable and a dangerous fire hazard. **(32)**

It is common to find polystyrene manufacturers claiming that they use CO2 as a blowing agent rather than pentane. Pentane has a distinctive sickly sweet smell which can be detected even after most of the pentane has off gassed. Styrene is one of the most dangerous chemicals in widespread use in buildings but manufacturers will claim that it is non-toxic.

> As early as the 1980s, The US Environmental Protection Agency demonstrated that styrene – the molecular building block of all polystyrene –was present in 100% of the samples of human fat that they collected from all 48 states in the continental United States. The EPA recognises numerous dangers that styrene poses to the central nervous system, and exposure to styrene can cause headaches, fatigue, Dizziness, confusion, drowsiness, malaise and difficulty in concentrating. Styrene is considered a possible human carcinogen by the World Health Organisation's International Agency for Research on Cancer.
>
> Styrene also appears to mimic oestrogens in the body, disrupting normal hormone functions, and possibly contributing to thyroid problems, menstrual irregularities, and other hormone-related problems, as well as breast cancer and prostate cancer. Chronic exposure to high levels of styrene can cause liver damage and nerve tissue damage. These effects can be especially pronounced in foetuses and young children.
>
> The gases that manufacturers pump into polystyrene to turn it into "foamed polystyrene" can also be hazardous. **(33)**

Polystyrene particles are found almost everywhere, and waste polystyrene is one of the biggest contaminants found in plastic waste in oceans. Polystyrene particle is easily ingested and can have serious health effects.

> We determined that PS particles were potential immune stimulants that induced cytokine and chemokine production in a size-dependent and concentration-dependent manner. **(34)**

Increasingly polystyrene insulation contains graphite powder which is claimed to improve its energy efficiency and reduce VOC emissions. Information on the health impact of the graphite powder could not be found.

Foam insulations

Petrochemical-based foam insulations are generally seen as high-performance solutions to insulate buildings, though the performance of various formulations varies a great deal. All foam insulations whether they be Polyisocyanurate (PIR), Polyurethane (PUR) or expanded polystyrene (EPS) contain a cocktail of volatile chemicals, including flame retardants and hazardous substances, whether they be styrenes or iso-cyanates.

Ambitious claims are made by manufacturers for the thermal performance of lightweight foam insulations, but these are often exaggerated, and may not even be true. Insulation companies were accused of making false claims about the fire performance of their products during the Grenfell Inquiry and this must also raise doubts about the claims of thermal performance.

> Celotex, which made the bulk of the combustible foam insulation used, displayed a "widespread culture … of ignoring compliance," which included distorting a full-scale fire test of its materials, the inquiry heard.
>
> Kingspan, which made the rest of the insulation, carried out tests that involved either "concealing components in a manner designed to facilitate a pass and/or using materials that were not as described in the test reports," it was claimed. Internal emails from the firm revealed it knew it was "dodgy" for it to advertise that its material could be used on tall buildings above 18 metres. **(35)**

It is not uncommon for questionable claims to be made about insulation performance, but while these may be correct when based on hot box tests in the laboratory, products can also shrink and deteriorate over time in buildings. Examples have been found of foam insulations that have shrunk.

> An insulation seller will pay a $104,257 civil penalty to settle charges that it made false and misleading performance claims about its insulation product. **(36)**

> A building company has sued for more than €2m damages over allegedly defective insulation boards which, it claims, caused floors in houses built by it to sink. Tallaght-based Kelland Homes says extensive remedial works were required to about 58 houses at a development at Elder Heath, Kiltipper, Tallaght. It has sued Ballytherm Ltd; Clondalkin Builders Providers Ltd, and James McMahon (Dublin) Ltd, who sold the insulation boards to Kelland; and a Dutch company, Covestro B.V, whose allegedly defective product was used in the manufacture of the boards. **(37), (38)**

It is possible that foam boards that shrink have not been properly cured and so chemicals may have continued to offgas at high levels even after the materials were placed in the building.

Counties Cavan and Monaghan in the Republic of Ireland are a significant centre for the reprocessing of chemicals, manufactured elsewhere in Europe, USA and China, into foam insulation products. However, in 2017, the supply of the raw material, MDI dried up, due to the closure of MDI production facilities in Europe, because of pollution problems and there were also two major fires at the BASF plant in Ludwigshavn.

> Irish insulation manufacturers have been forced to slash supply of some product ranges as a leading European supplier of a key ingredient experiences production problems, at a time when builders are passing on rising raw material costs to customers. Kingspan warned late last month that global shortages of the key chemical used to make rigid insulation panels, methylene diphenyl diisocyante (MDI), has been driving up prices for the material this year. **(39)**

> Just weeks after its spin-off from Bayer, German engineering plastics producer Covestro has announced its first streamlining measure, saying it plans to close its 170,000 t/y MDI plant at Tarragona, Spain, by the end of 2107. Around 170 jobs will be eliminated in the move. **(40)**

The Flix water reservoir, in Tarragona, has been receiving the impacts of the toxic chemical industry for more than a century. More than 700,000 cubic meters of toxic waste containing substances such as mercury, cadmium and other toxic organochlorine components (such as hexachlorobenzene, PCBs or DDT) have been dumped by the Ecros company to the Ebro river from Flix. **(41)**

As Ercros has been reporting in its annual accounts, the European Union has banned from 11 December 2017, the use of mercury technology in the production of chlorine and caustic soda, a fact that coincides in time with the decision announced by Covestro (formerly Bayer) to close its MDI plant. Due to Covestro is the largest consumer of chlorine from Vila-seca I factory, the closure of its plant makes unnecessary the investment that Ercros should undertake to replace the mercury technology, which currently accounts for 100% of the chlorine production capacity at the factory in Flix and 70% of the capacity at Vila-seca I factory. **(42)**

www.ercros.es/index.php?option=com_content&view=article&id=1183:estimated-impact-to-ercros-of-banning-the-use-of-mercury-technology-and-covestro-plant-shutdown-from-december-2017&catid=44:relevant-events&Itemid=101&lang=en

What does pollution from factory production have to do with indoor air quality? High levels of pollution and hazards from the production of the raw materials for foam insulations correlates with possible emissions from these materials when used in buildings. Chemical pollution from the production of chemical building materials should be seen as parallel with chemical pollution in indoor air quality.

There may be insufficient data available about emissions from foam insulation materials in buildings, even though it is known that they contain a cocktail of dangerous chemicals, particularly flame retardants because manufacturers claim that their products are stable and emissions levels are very low. They often make patently absurd claims.

PU insulation (PUR/PIR) is produced by a reaction of diisocyanates (MDI) with polyols or themselves to create the solid PUR and/or PIR cell structure. MDI (methylene diphenyl diisocyanate) is a respiratory sensitizer and labelled R40(H351) – suspected of causing cancer. The MDI is chemically consumed during the foaming process and hence not present in the final rigid foam product. Third parties tested a range of PU insulation products using recognised test methods in order to verify whether any MDI emissions were detectable. They all confirm that there are no MDI emissions from these products. PU insulation is considered a very low emission product. In fact, emissions from PU products are well below those of most other insulation products. In particular, natural insulants can have VOC emission levels more than 100 times higher than those of PU. Very importantly, no carcinogenic, mutagenic or reprotoxic substances were detected in any of the emission tests on PU foam. **(43)**

A range of isocyanates are used in products including polymeric MDI (PMDI) and HDI, TDI, IPDI and MIT. They are skin irritants and can cause asthma inflammation, dermatitis and respiratory sensitisation. Surprisingly, considering the widespread use of Isocyanates, epidemiological research into health effects has been limited. **(44)**

Isocyanate production is highly dangerous and pollution emissions of real concern. The release of Methyl Isocyanate from the explosion and chemical leak in Bhopal in India in 1984, immediately killing 3,800 people but also affecting the long-term health of many thousands more, is perhaps the most infamous example of the dangers of isocyanates. **(45)**

Isocyanates are derived from Phosgene, better known as mustard gas, and is widely used by chemical companies today, particularly in the production of Toluene di-isocyanate and many other hazardous chemicals which find their way into insulation and building products. They are made from Chlorine, activated carbon and carbon monoxide. The main producers of phosgene today are Dow Chemicals, Covestro, and BASF in Texas and Louisiana, China, Japan and South Korea. **(46)**

Spray foam

One of the most worrying forms of foam insulation is spray foam. Heavily advertised, with high-pressure salesmen scouring the country, to persuade unsuspecting householders to pay a lot of money for foam insulation to be sprayed into roofs, lofts and sometimes walls.

> …some of the chemicals used to make spray foam installation are known to be hazardous to human health, according to the Environmental Protection Agency (EPA). Spray foam installers must wear protective gear while spraying. According to the EPA, "homeowners who are exposed to isocyanates and other spray foam chemicals in vapours, aerosols, and dust during or after the installation process "run the risk of developing asthma, sensitization, lung damage, other respiratory and breathing problems, and skin and eye irritation."

In the USA and Canada numerous class action lawsuits have led to the collapse of companies manufacturing and installing spray foam and lawyers have been seeking victims of spray foam companies. **(47)**, **(48)**, **(49)**

Spray foam is made by combining MDI with flame retardants and a blowing agent. While this toxic combination is used in many foam products, they have at least had time to cure and off gas before arriving on site, whereas this curing happens in the house for spray foam, with people sometimes continuing to live in the premises while the work is done, being told to vacate for only 24 hours thus receiving a much higher dose of the chemicals as they initially offgas.

The installers are meant to wear protective masks and clothing but there have been a small number of deaths of installers in the USA. There seems to have been less concern in the UK about the health of householders where spray foam has been installed, and at least one of the companies is using products from one of the bankrupt US companies.

Local authority and energy efficiency grant organisations have been promoting the use of spray foam and there is even a company that uses a small robot to spray foam underneath floorboards in basements. Banks and building societies have taken steps to limit the application of spray foam due to concerns about a reduction in the value of the property. Many homeowners have found that they cannot sell or re-mortgage their house when spray foam has been installed.

> Spray foam insulation might: reduce air circulation and ventilation within a roof space; lead to dampness and condensation on the underside of a roof because it forms an air barrier and stops moisture from escaping. Placing timber-framed roofs at risk of decay. Spray foam insulation can also be difficult and costly to remove. According to the website Checkatrade, the cost of removing spray foam insulation from the roof of a three-bedroom detached house is around £3,200 (or £40 per square metre). **(50)**

This estimate of the cost of removing spray foam is a serious underestimate, as often houses have to be completely re-roofed, and in cases where spray foam has been removed from cavity walls, the whole inner leaf of the house has had to be taken down. It says something about UK culture that spray foam is not removed due to its alleged toxicity and effects on health (which has been the driving force in the USA) but because it has reduced the mortgage value of a house!

Spray foam has also been cited as a fire hazard in the USA following three houses fires in Massachusetts.

> Fire investigators suspect that a fire that destroyed a $5 million home in Woods Hole, Mass., was ignited when excess heat was generated by the exothermic reaction that occurs during the installation of spray polyurethane foam. **(51)**

Glues, sealants and fillers

Polyurethane (PUR) is widely used in adhesives and fillers. While mechanical fixings are still used in construction, glueing has become more common. So extensive is the use of PUR adhesives that indoor exposure can reach safe threshold values. Many of the products used by Do it Yourself enthusiasts (DIYers) emit the highly dangerous isocyanates at very high levels. Wirts et al have measured emissions and curing rates but provide very little information on the health effects. **(52)**, **(53)**

UFFI urea formaldehyde foam

Despite the severe hazards associated with urea formaldehyde it has not been banned from use in the buildings in the UK. While it was used in a cavity construction in houses in the 50s and 60s, when discovered it is usually extracted. Often the material has decayed and is found as a brown sludge at the bottom of the cavity. It was used in social housing at one time but records of how many affected houses cannot be found. There must remain concerns that the health of occupants of houses where UFFI was used have been damaged as it may still be present in many houses.

> Insulating a home with urea formaldehyde foam can lead to severe health problems due to poisoning from formaldehyde gas. Respiratory problems, allergies, memory loss, and mental problems can result from exposure to foam insulation fumes. Research is now under way at the Chemical Industry Inst., Univ. of Washington, and other institutions to learn more about the health effects of formaldehyde foam and to develop possible remedies to these problems. Several states are either banning or controlling the use of this type of home insulation. **(54)**

Despite the knowledge of the risks of UFFI it was used to insulate under the floor of Flash Ley primary school in Staffordshire in 2015. The work was carried out along with other repairs during the summer vacation but when the children returned to school 20 pupils immediately became ill and the school had to be evacuated and shut down for a year. The local authority claimed that the cost of dealing with the problem was over a million pounds.

> Flash Ley School (Hawksmoor Rd, Stafford ST17 9DR) was evacuated in October 2015 after formaldehyde was discovered with pupils returning this January2017 after being taught elsewhere. **(55)**

It cost the local authority £1,113,159.14 to rectify the problems and remove the hazardous material, but the local authority refused to provide any details, as they said it was commercially confidential! The work involved the use of a product called Benefil which was normally used for filling old pipes like gas mains that have been taken out of use, but a high-density version of the material is also used to fill old mine shafts. **(56)**

> The formaldehyde was released as a result of works that took place last summer to fill ducts in the floors throughout the school to provide additional support. Benefil issued a statement which said: "Benefil provided and installed the product at Flash Ley School over the summer of 2015, as subcontracted by Entrust (owned by CAPITA) on behalf of Staffordshire County Council." **(57), (58)**

Given the reluctance of the various official bodies involved to share any information, it has not been possible to discover who thought it was appropriate to use such a material in a way that it could off-gas into the school. No trace could be found of this material being used elsewhere in schools, but it is worrying that such a dangerous material could be approved for use in this one. Capita, who commissioned this work in a partnership with Staffordshire council, offers Property and Facilities management services. **(59), (60)**

As far as could be established there have not been any follow-up studies to monitor the health of the children who were moved back into the school. Public Health England seemed remarkably complacent about the incident, dismissing possible acute effects from what may have been very high-level but short-term emissions. It was not possible to find any scientific research carried out by Public Health England to justify its questionable views about short-term exposure whereas the US EPA seems much more concerned.

Public Health England did respond to a freedom of information request about whether there would be any long-term health effects of the exposure to formaldehyde and stated:

> Public Health England (PHE) has reviewed the scientific evidence to find out if there are any longer-term risks to health from this incident. Their advice is that children and staff at the school who are currently well are unlikely to have suffered any impacts to their long-term health.
>
> There are limited data available on the long-term effects of acute or short-term exposure to formaldehyde. There is some evidence of a reduction in lung function and exacerbation of asthma symptoms however the majority of studies show no effect on lung function in either asthmatics or non-asthmatics.
>
> Chronic occupational exposure to formaldehyde has been reported to cause continuous eye irritation, rhinitis and nasal lesions. These effects are as a result of continuous exposure for very long periods of time so are unlikely to be relevant to the situation at Flash Ley Primary School.
>
> Formaldehyde has been classified as a human carcinogen (group 1) by the International Agency for Research on Cancer (IARC). There is evidence that formaldehyde causes nasal tumours (and leukaemia in industrial workers exposed over long periods of time (years). However, exposure over short periods of time is very unlikely to have the same level of risk.
>
> Any effects from exposure to formaldehyde would have been short lived and any ill effects very short term. It is highly unlikely anyone exposed will have long lasting health effects therefore follow up is not required. We are not aware of any long term adverse health effects in children due to exposure to formaldehyde related to this incident. Any information with respect to clinical management and treatment of individuals affected would be at the discretion of their GP or NHS Trust. We do not hold or have any information pertaining this. **(61)**

In 1987, the U.S. Environmental Protection Agency (EPA) classified formaldehyde as a probable human carcinogen under conditions of unusually high or prolonged exposure (*1*). Since that time, some studies of humans have suggested that formaldehyde exposure is associated with certain types of cancer. The International Agency for Research on Cancer (IARC) classifies formaldehyde as a human carcinogen (*2*). In 2011, the National Toxicology Program, an interagency program of the Department of Health and Human Services, named formaldehyde as a known human carcinogen in its 12th Report on Carcinogens (*3*). **(62)**

Plasterboard and gypsum materials

The use of plasterboard, or drywall as it is known in the USA, is so commonplace that it seems unusual to include it here as a hazardous material. Plasterboard is relatively cheap and is widely used as dry lining, internal walls or as the internal lining of timber frame construction. Plasterboard is compressed gypsum and other materials lined with paper and can easily be plastered with gypsum plaster. Most plasterboard is reckoned to have a half hour fire resistance and sometimes two layers are used to give one hour. When plastered, the boards have limited vapour permeability, so some moisture vapour might be able to pass through in either direction, but this means that chemical emissions can also migrate through plasterboard to contaminate internal air. Once the walls are painted with plastic paint, however, the vapour permeability is reduced.

Most gypsum boards are a standard product with British Gypsum, owned by Saint Gobain enjoying a near monopoly in the UK and Ireland along with their main competitor, Knauf. There are alternative boards from Germany, also made with gypsum but claiming better environmental characteristics and alternative boards made with natural materials and clay are being developed. The gypsum content has for many years been made from power station and other combustion waste, known as DSG (Desulpherised Gypsum). As a recycled material, this might seem positive but traces of heavy metals and sulphur dioxide may be present and there was a famous scandal in the USA where "drywall" boards, imported from China, retained high levels of sulphur, and various class actions were pursued in what became known as the drywall scandal. Survivors of Hurricane Katrina were offered houses built using this drywall but refused to move in as the houses smelled strongly of bad eggs (Sulphur Dioxide). Metal electrical fittings corroded within a short space of time. Various oppress reports claimed that houses built by the Christian Charity Habitat for Humanity had used the defective drywall. **(63)**, **(64)**, **(65)**, **(66)**

In recent years, the source of recycled materials for DSG has reduced due to the reduction in coal-fired power stations and blast furnaces. This has led to companies re-opening natural gypsum mines, so that possibly there will be less hazardous chemicals in plasterboard and gypsum plasters, but it may be that recycled materials will be mixed with natural gypsum.

Efforts have been made by the industry to recycle waste plasterboard, most building sites will have masses of plasterboard offcuts, but plants where gypsum boards are recycled can be dangerous places with Health and Safety Executive have levied fines for risks to the health of the workers in recycling plants. Dust from plasterboard is a serious health hazard. **(67)**

A typical health and safety data sheet will state that Gypsum wallboards are made from Calcium Sulphate Dihydrate (CASO4) encased in paper liners but may also include minor amounts of quartz, small quantities of chopped glass fibre, micro silica, vermiculite starch, foam and dispersants. Some products may include fungicides. A silicon hydrophobic agent may be used to make boards more water resistant. DSG can contain trace elements of metals, particularly mercury and in 2010, the U.S. Environmental Protection Agency (EPA) released a study of total content and leaching values of heavy metals in synthetic gypsum, which found that these chemicals could have leaching values of up to 550 times the level for safe drinking water. Total content, on the other hand, never exceeded a measurement. **(68)**

The manufacture of desulpherised and recycled materials requires environmental permits such as in Portbury Somerset which involve chimney stacks with emission concerns. **(69)**

It has not been easy to track information about which products are treated with biocides or fungicides. A study in Finland confirms that mould spores can grow on plasterboard. Spores collected from plasterboards "were toxic to the macrophages. The biocide added to the core did not reduce the growth; in fact, the spores collected from that board evoked the highest cytotoxicity" **(70).**

Gypsum is mined throughout the world including Mauritius, Tunisia and other parts of Africa as well as Ireland where a Gaelic football ground and buildings recently disappeared into a sink hole and other subsidence has occurred in Counties Cavan and Monaghan in Ireland, with extensive media coverage **(71)**, where mines had been extended. **(72)**

Multi-national cement companies play a big role in gypsum mining and there are many concerns about environmental and health damage from mining such as at the Afsana Batt Gypsum Extraction in Kashmir. **(73)**

While plasterboard companies can be a little shy of admitting their use biocides or fungicides, these may be used only in the paper lining, and thus not appear in data sheets, a range of companies advertise chemicals for use in plasterboard and plasters **(74).**

Thiabendazol, Azoxystrobin and Fludioxonil are fungicides for the protection of building materials for indoor applications such as gypsum board. Ethylenediamine is also used. Icopal is a fungicide used in a wide variety of building materials. Many plasters contain salt-inhibiting chemicals and are widely used in renovation of historic buildings, plasters can be made with portland cement with Calcium dihydroxide and crystalline silica. Borax is added to some plasters. **(75)**

Gypsum plasters and plasterboards are generally presented as being non-hazardous but cannot limit or eliminate VOC emissions from other products such as composite timber boards

> ...plasterboards have only a small effect on TVOC levels, but are not capable of reducing VOC emissions from building products. **(76)**

In many building projects, plasterboards are glued to other boards or foam insulation batts creating a complex and possibly hazardous combination known as dry lining.

> Plasterboard is used extensively as a dry lining board in timber kits. These boards differ in their use of adhesives but it should be understood that VOCs may be emitted from the binding adhesive used for fixing papers to the plaster substrates. Absorbed solvents from the coatings of plasterboard can release steady concentrations of VOCs into indoor atmospheres over long periods of time. **(77)**

Concrete, masonry, mortars and floor screeds

It's hard to build a building without using cement and concrete though, it is possible to minimise its use. It is often said that if the cement industry were a country, it would be the third-largest emitter of carbon dioxide in the world after the USA and China and yet the cement and concrete industry continues to produce advertising material saying it is a sustainable product. **(78)** www.sustainableconcrete.org.uk/

The production of cement uses a lot of energy through its firing of basic materials, lime and clay, in kilns. But increasingly cement-based products, which arrive on site, also contain a range of added chemicals. Walling materials, external renders, flooring screeds tiling and adhesives use added chemicals such as styrenes, blast furnace slag, ferrous metals, 2,2,4-trimethyl-1,3-pentanediol monoisobutyrate, (mixture of: 5-chloro-2-methyl-2H-isothiazol-3-one [EC no. 247-500-7] and 2-methyl-2H -isothiazol-3-one [EC no. 220-239-6] (3:1) 1,2-benzisothiazol-3(2H)-one, carbon black, magnetite and gypsum magnetite, crystalline silica.)

A scan of a range of health and safety data sheets for composite cement-based products in common use will find these and many other additives, too many to list here. In the past urea-based anti-freeze chemicals were used in concrete which then generated ammonia releases into buildings. **(79)**

Contamination of concrete by heavy metals, even uranium, have been tracked, but there seems to have been much less research on these topics in recent times. Other studies have identified ammonia and VOC emissions in indoor air. **(80)**

A wide range of so-called eco cement and eco concrete products are now available with architects and engineers convinced that these have a lower environmental impact, as less energy is used to produce them than ordinary portland cement. Ground Granulated Blast Furnace Slag (GGBS) contains ferrous metals but data sheets provide little information as to other contents. There remain concerns about hazardous emissions from incineration of materials which are then used in GGBS. **(81)**

The effect of the many modern additives to cement and concrete products has yet to be fully explored in terms of indoor air emissions. It may be that the levels of risk are low, but using the precautionary principle, when installing a floor screed, it might be better to use simple cement and sand rather than a range of composite products that have many chemical additives to help them flow or dry out, such as Styrene (Flowchem) Carbon Black and Magnetite (Isocrete) (Flowcrete). Latex is added (Ronascreed). Ardex is a water-based PVA primer and bonding agent which is often used on porous surfaces before the application of plasters or cement renders. PVA is widely used by builders for all sorts of purposes. Ardex according to its safety data sheet P 51 (8/22/2016 Revision date: 9/10/2021) includes 2,2,4-trimethyl-1,3-pentanediol monoisobutyrate, mixture of: 5-chloro-2-methyl-2H-isothiazol-3-one [EC no. 247-500-7] and 2-methyl-2H -isothiazol-3-one [EC no. 220-239-6] (3:1) and 1,2-benzisothiazol-3(2H)-one.

Some concrete mixes and plasters have additives to act as a barrier to x-rays. **(82)**
Many problems exist as a result of experiments in innovative forms of concrete such as autoclaved blocks and the addition of pyrites and mica to concrete (see Chapter 9). Many thousands of houses and schools are falling apart thanks to the use of these materials. The addition of mica, for example, has been shown to reduce mix cohesiveness and increased porosity in walls, allowing greater water penetration. It is an indication of how the industry is happy to use untried and risky composite solutions for building without considering the long-term consequences and are unlikely to have considered possible indoor air pollution risks.

Composite timber boards

A major source of emissions in buildings is from a rage of composite timber products. Some timber boards are part of the structure of a building, while others are part of finishes such as kitchen units and furniture.

A great deal of modern construction by housing developers uses timber frames of various kinds, usually prefabricated. These timber panel buildings use sheathing boards of OSB (Oriented Strand-Board) which is made with significant quantities of glues. The timber frame is also wrapped in plastic air tightness membranes, and various synthetic insulation materials are lined with gypsum board internally. External cladding can be of brick and other materials. This is not a healthy use of timber and some of these forms of construction can be highly flammable. A typical health and safety data sheet for one of the best known OSB boards contains the somewhat bizarre statement "Non-flammable at room temperature, **but will burn."** Data sheets also contain the statement,

> OSB in panel form is unlikely to give rise to any toxicological effects; however, health risks may arise from dust and moulds associated with poor processing, handling or storage practices. **(83)**

OSB contains significant amounts of adhesives but the chemical composition of these is rarely disclosed, other than occasional references to Formaldehyde. The boards also receive a "coating" which is usually not specified either:

> Formaldehyde has a WEL of 2.5 mg/m3 (8 hour TWA) and a Short Term Exposure Limit (STEL) also 2.5 mg/m3 (15 minute exposure).Formaldehyde vapour can irritate the eyes and nasal linings. Formaldehyde class for OSB is class E1 – less than or equal to 8 mg/100g (0.008 %) of board as per BS EN 13986:2004 Annex B. **(84)**

A leading brand of OSB is advertised as having zero **added** formaldehyde though the MDI that is used is also hazardous.

> Unlike traditional oriented strand board (OSB) containing urea-formaldehyde (UF) or melamine-urea-formaldehyde (MUF) binders, the Sterling OSB Zero family from Norbord uses a methylene diphenyl diisocyanate (poly-urethane MDI) resin to bind the thousands of strands that make up each board. **(85)**

Stating that OSB is free form formaldehyde clearly has the intention to make it sound healthier and greener but according to the US EPA, Diisocyanates are also dangerous.

> ...general population exposures, particularly in or around buildings, including homes and schools. Diisocyanates are well known dermal and inhalation sensitizers in the workplace and have been documented to cause asthma, lung damage, and in severe cases, fatal reactions. **(86)**

Natural timber also contains chemicals such as formaldehyde, and this is sometimes used as an excuse for formaldehyde emissions. Some products such as kitchen units are advertised as low formaldehyde, but this means that other adhesives such as PUR have been used, which can also be responsible for chemical emissions. It is possible to source kitchen units made from timber-free metal alloys though these may have been painted with paints that might also produce emissions. Many kitchen units are coated with acrylic finishes.

Composite timber boards also include plywood (hardwood), plywood with composite cores, particle board (PB), medium density fibreboard (MDF) and various laminates. The USA has formaldehyde limits for composite wood products which became law under President Obama in 2010, and there are also legal limits in Canada.

It is generally claimed that boards produced today are "low emission" and yet formaldehyde emissions will usually be found in most modern homes, often above safe limits. It's unlikely this can be blamed on the occupants use of cosmetics and cleaning materials, even though formaldehyde is used in some shampoos. MDF is made with synthetic resin adhesives and many people are aware of dangers from MDF dust, particularly with sawing or sanding products. The UK Health and Safety Executive provides some limited guidance but there do not appear to be legal limits in the UK like the USA. Despite the known carcinogenic effects of formaldehyde and the hazards from other adhesives, in the UK there is very little control. **(87)**

Solid timber

There is growing popularity for solid timber construction, especially among architects. This involves using structural timber boards referred to as Cross Laminated timber (CLT) or mass timber. Multi-storey buildings are now being constructed using prefabricated solid timber panels for walls, floors and ceilings. The Mjøstårnet 81-metre-high tower in Brumunnddal, in Norway, is currently the tallest solid timber building in Europe. There are many solid timber buildings in the UK including a few multi-storey housing projects. Suppliers of solid timber systems are rarely forthcoming about adhesives, but so far there is little, if any, evidence of concerns about emissions from solid timber construction.

Doors: wood and glass fibre

Doors made from composite timber products are increasingly popular with housing developers and DIYers. There are a vast range of products available but the most widely used are often pre-finished with wood textured paint and may not even be made out of wood but glass fibre (or fibreglass as is the term often used in the USA).

One of the biggest manufacturers of composite doors is based in Ireland. Reviewing a range of websites it proved to be very difficult to get access to fire safety and health and safety data sheets. The only one easily accessible was from an Australian website. The company in Ireland was contacted and asked for health and safety data sheets. They have not yet replied.

Many doors are made with a wood veneer on a particle board, fibreboard or hardboard core and some are claimed to be fire resistant with "mineral" cores. Some doors have hollow cores and are very light. Some have high pressure laminate finishes and some are even lead lined. As these products can contain paraffin wax, phenolic resin and maybe linseed oil they are potential sources of emissions. The processing of the composite timber involves the release of natural chemicals within the timber as well as added glues.

There have been environmental concerns in the past (1999) about discharges in Ireland into the River Shannon of formaldehyde, a carcinogen. Without access to health and safety data sheets it is impossible to tell whether formaldehyde is still used in the manufacturing process of composite doors and no recent evidence of discharge concerns since 1999 could be found.

Some useful images can be found showing the difference between solid wood, composite and engineered wood doors by Direct Doors who warn about many products being sold as solid wood doors when they are not. **(88)**

There are organisations claiming to supply "non-toxic" doors, but they still advocate the use of epoxy glues as well as other less toxic products. **(89)**

Air tightness membranes and tapes

Most air tightness membranes are made from polyethylene plastic. It is possible to achieve airtightness barriers using insulation boards, timber sheathing boards, building paper and wet materials such as plasters. However, many advocates of zero carbon buildings prefer to use plastic membranes, with the joints sealed with air tightness tapes. Polyethylene achieved notoriety in the Grenfell fire as it was used inappropriately in the external cladding panels, which combusted at a relatively low temperature (90 degrees C), and so the widespread use of such a flammable material inside buildings, has to be a cause for concern. The membranes are usually referred to as breathable or vapour permeable and there are some products that are referred to as "intelligent" membranes.

Roofs, under the slates, tiles or metal sheets, were constructed for many years with bituminous membranes as a secondary waterproof barrier, but these have been replaced with plastic membranes. The theory is that the membranes will keep water out but allow internal vapour to escape through the membrane, thus avoiding the risk of condensation. It is very difficult to find independent evaluation of the efficacy of vapour being able to escape through these membranes and those studies which could be found relied on computer models rather than actual tests of performance in buildings. Much of the literature seemed pre-occupied with the tapes as they seem to inhibit the vapour permeability, and the limited literature available was published by manufacturers rather than independent researchers. **(90)**

In order to reduce the risks of fire, most membranes contain fire-retardant chemicals, but a number of leading brands do not disclose these chemical constituents in their health and safety data sheets, simply claiming that the product is not hazardous. Indeed, many manufacturers did not seem to publish safety data sheets at all. After searching through many that do not disclose their chemical composition, one was found that admitted to the use of phenol formaldehyde. Other chemicals might include phosphorous, melamine, nitrogen, inorganic hydroxides, boron and silicon but the only literature available is for polyethylene in general. **(91)**

Given the widespread use of polyethylene membranes in buildings, much more needs to be known about the flammability of these products and also whether there is a danger of flame-retardant emissions. An event on breather membranes was organised by the RIBA entitled "Are your wall membranes non-combustible and smoke-blocking?" and the resulting published RIBA article implies that many membranes commonly in use are combustible and it calls for non-combustible membranes, but doesn't explain how these might be made. The organiser Allan Hurdle of AKH services appears to be a consultant to one of the manufacturers promoted at the event, Stamisol. It was possible to find a health and safety data sheet for Stamisol which shows that the product contains Trimethoxyvinylsitane and Diisnonyl phthalate. Trimethoxyvinylsitane was evaluated under the REACH directive and concern for its mutagenic potential was identified. **(92)**, **(93)**, **(94)**

There is little doubt that PFAS are used in breather membranes and a number of companies, due to widespread public concern, are now beginning to advertise breathable materials (especially for clothing) as PFA free. So far, no PFA-free membranes for buildings could be found.

Air tightness tapes that are used with membranes, but are also applied to various boards, are self-adhesive and some products claim they are solvent, formaldehyde and emulsifier free. The tapes are generally made from PVC, glass fibre and other plastics. One rare data sheet shows a tape containing 1,1-(Ethane-1,2-diyl) bis[pentabromobenzene] 84852-53-9 10-20 Acrylic Adhesive – 30-40 Polyethylene liner.

"Natural insulations" that use hazardous chemicals

There are a wide range of materials, including insulations, claiming to be "natural" and "eco-friendly," available today. There are no set legal standards for the use of the term eco-friendly (or even natural) and the term is often interpreted as meaning energy saving or energy efficient, even if it is made from hazardous, toxic and planet damaging chemicals.

Eco or natural building products can include recycled cellulose (newspaper), sheep's wool, wood fibre and cork composites. They may use chemical additives that may be bad for indoor air quality. Chemicals are added as adhesives and binders, to act as flame retardants and preservatives against mould, pests and moths, for instance. In most cases, there are chemical-free versions of these products.

The main problem in terms of indoor air quality is caused by the use of Borax, Boric Acid or Sodium Borate. Some environmental award winning architects are fully committed to using insulations heavily dosed with borax, and have been using these materials for many years. The multi-national company Rio Tinto are said to control the majority of Borates mining on a global scale. It doesn't take long on the internet to discover many articles criticising the environmental record of Rio Tinto which produces aluminium, copper, diamonds, gold, industrial minerals, titanium dioxide and salt along with iron ore. **(95)**

Recycled cellulose insulation uses a combination of chemicals including Boric Acid and Magnesium Sulphate. Manufacturers are not always very open about the proportion of these chemical additives but there are some companies who offer "borate free" or "reduced borate" products. Boric acid, Boric Oxide, Sodium Borate and Sodium Perborate, hereafter referred to as "borates," are generally registered under the REACH directive. The EU has battled with the borates industry for many years with attempts to ban its use, due to its suspected negative health effects, including concerns about the use of borates in cosmetics. There is new classification of some boron compounds as mutagenic and/or toxic to reproduction according to the Commission Regulation 790/20091. Borates are seen as a potential slow release chemical and Borax ($Na_2B_4O_2(H_2O)_{10}$), a low toxicity mineral with insecticidal, fungicidal and herbicidal properties.

> Other release to the environment of this substance is likely to occur from: indoor use in long-life materials with low release rate (e.g. flooring, furniture, toys, construction materials. **(96), (97), (98)**

Borates are widely used, not just in insulations but composite cork products, composite wood products, paper and cardboard and many others. A number of green organisations have argued that borates are not harmful and that it cannot be ingested by humans but the various forms of borates have been listed by the EU as persistent organic pollutants. Borates are widely used in agriculture and even in fireworks to create a green colour.

Direct experience of borates in insulation, involved a small conference at a sheep wool insulation factory in North Wales, (no longer in existence), where some participants had such severe asthmatic reactions to the borate dust in the air in the building that the event had to move to a nearby hotel. When installing borate dosed sheep wool insulation in a loft, the installers suffered from eye infections. Installers of cellulose insulation are meant to wear protective clothing, eye protection and masks, but it is not clear whether the borates involved remains as volatile once encapsulated within lofts or walls. Borax is often combined with other chemicals such as Aluminium Sulphate and Phenol Formaldehyde and are used in man-made mineral fibres. **(99)**

Borates are widely used in the environment as a household pesticides and finds its way into food and water. It was not possible to find any indoor air quality emissions studies that have detected borate emissions from buildings where insulation has been used. However, concerns remain high and it is possible to obtain a range of building products that do not use this.

Flame retardants

> Over-reliance on flame retardants inevitably creates serious and largely unconsidered problems for future generations. Over time, flame retardants are released into homes, offices, buildings, and vehicles through a combination of volatilization, abrasion of fibres and particles from treated fabrics, and as a result of foam degradation. Because products are not labelled or bar-coded, it is not known what flame retardants are used in foams and, even if it were, a flame retardant considered safe today may be found not to be in the future. Currently it takes years or even decades to restrict chemicals under both REACH and the Stockholm convention (POPs). When this happens, there is no easy and economical way of identifying and removing the affected items, especially when they are used in building insulation. It would be much more sensible to use materials that are intrinsically fire resistant and safe. **(100)**

Emissions of flame retardants in insulations can affect indoor air, particularly where dry lining is used, bringing the materials very close to occupants. The UK is one of the highest users of flame retardants in the world and this led to an enquiry by the Westminster parliament environmental audit committee in 2019.

While flame retardants are intended to control fire and smoke toxicity in buildings, they have the negative effect of introducing yet more hazardous chemicals into indoor air. Flame retardants may not always be effective, and rather ironically, when the products containing flame retardants burn, the retardants themselves may exacerbate the yields of toxic gases from burning foam, one of the principal causes of deaths in fires. Toxic smoke can lead to death and serious illness among people trying to escape fires and the chemicals can have a long term and serious impact on fire fighters.

Emissions from flame retardants in insulations can affect indoor air, particularly where dry lining is used, bringing the materials very close to occupants. The UK is said to be one of the highest users of flame retardants in the world and this led to an enquiry by the Westminster parliament environmental audit committee in 2019. **(101)**

Despite strong recommendations from the Environmental Audit committee, there is little evidence of any action that has been taken to reduce the use of flame retardants in the UK. Whereas major changes to ban flame retardants in some products have been taken in California and the Irish Government issued a consultation into their use in furniture in 2020 but flame retardants still have few restrictions.

An understanding of the dangers of flame retardants is not new. In 2007 Bellingham and Sharpe raised concerns about Polybrominated Diphenyl Ethers (PBDEs), in a paper which drew attention to the dangers for pregnant woman of a wide range of chemical exposures.

> An "impact paper" published today by the UK's Royal College of Obstetricians and Gynaecologists (RCOG) has drawn heavy criticism for suggesting that pregnant women should avoid exposure to common environmental chemicals, including BPA, despite conceding that the risks of such exposures are unknown and unlikely to be "truly harmful to most babies". The report is an opinion paper designed for health professionals but has attracted

> a large amount of media interest for suggesting that pregnant women avoid, among other things, new furniture, paint fumes and food from cans and plastic containers. **(102), (103)**

PBDEs are a class of halogenated flame retardants and are rarely used today, because they were phased out as POPs, but they still appear in some health and safety data sheets. The European Union banned the sale of two commercial mixtures of PBDEs, known as PentaBDE and OctaBDE, in concentrations higher than 0.1% by mass. A total ban has not been put in place and the withdrawal of PBDEs have been voluntary in some countries. Deca BDE is also listed as a POP. **(104)**

However, PBDEs are still widely found in the environment, particularly in sea going mammals, and they may well still be present inside buildings and furniture. **(105)**

PBDEs can affect neurodevelopment in children, including impaired cognitive development (comprehension, memory), impaired motor skills, increased impulsivity and decreased attention. And the US EPA has assigned a classification of "suggestive evidence of carcinogenic potential" for decaBDE. **(106)**

Manufacturers of insulation and other products used inside buildings change their fire-retardant formulations frequently, and even when these are used, they do not always appear on product data sheets. Organo-phosphates have been used in some insulations, as a replacement flame retardant. **(107), (108), (109)**

Little attention was paid to what happens to insulation materials in fires until the Grenfell disaster occurred. Given the high level of hazardous chemicals used in flame retardants it can be argued that there is a correlation between the chemical emission risk to indoor air quality and the toxicity of these materials in fires. Insufficient work has been done to measure emissions from flame retardants in insulation materials due to an assumption that they are not hazardous.

A comprehensive study of smoke toxicity was undertaken by Yates at the BRE in 2017. The report contains a wealth of information on fire safety and the dangers of insulation materials in fires. Yates refers to the toxicity of six insulation materials (glass wool, stone wool, expanded polystyrene foam, phenolic foam, polyurethane foam and polyisocyanurate foam) that were investigated under a range of fire conditions in the ISO/TS 19700 steady state tube furnace by Stec and Hull, the PIR insulation contributed more heat to the fire. In a Warrington fire test compared with mineral wool, the PIR also produced hydrogen cyanide. **(110), (111)**

Smoke toxicity is an unregulated area of materials certification and an opportunity to regulate this during revisions to the EU Construction Product Regulations was blocked by the chemical industry lobby. There is a lack of a clear definition of terminology in fire safety engineering, injury and death which has limited the collection of data to produce coherent fire statistics at EU level. The Modern Building Alliance, which represents most of the European plastic foam and related companies, could hardly disguise their delight that industry had managed to block the incorporation of smoke toxicity restrictions.

> The European Commission confirms that regulating the smoke toxicity from construction products would not be justified. **(112)**

While manufacturers may claim low emissions from plastic foam insulations, particularly those that contain isocyanates, the volatility of the chemicals involved is evidenced by the production of cyanide in fires. Exposure to chemicals released in fires, even if people survive the fire, is of great concern to firefighters where it is claimed that they have much higher risks of contracting cancer.

Table 2.2 Chemical contaminants in building materials

List of chemical contaminants widely used in building materials	*Potential health impacts*	*Use in building materials*
Lists derived from several documents and sources including: **(116)**, **(117)**, **(118)**, **(119)**		
Flame retardants, PBDEs and organophosphates	Cancer, neurological damages, hormone (endocrine) disruption, neuro development delays, obesity	Widely used in insulation, building products, furnishings
Chlorinated Tris flame retardants	Cancer	Widely used in insulation, building products, furnishings
Borates/Borax Sodium Borate Zinc Borate Sodium Tetraborate	Allergy Irritation of the nose and throat, increased nasal secretion, wheezing, coughing, pulmonary oedema, cardiac arrhythmias and coma	Used as a flame-retardant fungicide and insect repellent in insulation materials such as some sheep wool products, recycled newspaper insulation, glass fibre, paper-based composites, treated timber, concrete, ceramic glazing, TV screens
Vinyl, pthalates, methyl methacrylate	Effects on testosterone, hormone levels allergies and asthma. Pthalates are a semi-volatile organic compound. Asthma effects of greatest concern	Used in vinyl flooring and a wide range of plastic lining materials, edging and finishes
Formaldehyde	Irritation of nose and mouth, Cancer. One of the most pervasive hazardous chemicals in buildings. Category 1B carcinogen. Restrictions in REACH	Composite and engineered wood products, cabinets, counter tops mouldings, shelving, stair systems wall linings, composite structural timber like I beams, fabrics, glues, paints, caulks
Urea formaldehyde (UFFI)	Irritates eyes, skin and respiratory tract, coughing and choking can cause allergic reactions. Listed in REACH as a restricted substance. Bizarrely allowed in UK building regulations in cavity construction. Long history of attempts in Canada and USA to ban it	Thermosetting resin and polymer, used as an insulation and major use as an adhesive and a bulking agent in cosmetics
Polyethylene	Dermatitis, skin irritation, eye burning, asthma, hormone disruption, asthma. Highly flammable	Used to make vapour barriers (plastic wrap) where it is treated with flame retardants. Gained notoriety as insulation layer in aluminium cladding. Widely used as pipe insulation

(*Continued*)

Table 2.2 (Continued)

List of chemical contaminants widely used in building materials	*Potential health impacts*	*Use in building materials*
Polypropylene Contains Pthalates	Said to be safe as food container as it does not contain Bisphenol A, but chemicals can leach out. May include Bisphenol A when used in building materials	Wide variety of films and sheets and vapour barriers, adhesives and tapes and piping
Bisphenol A	Mainly of concern through use in food packaging but it can leach out and is found in high levels in most people. Is regarded as an endocrine disruptor	Found in many plastic products listed above
VOCs Petroleum products		
Benzene and Ethyl benzene	Drowsiness, dizziness, rapid or irregular heartbeat, headaches, tremors, confusion, unconsciousness. High risk from attached garages (Petrol) and heating systems Quite common VOC in indoor air tests but hard to assess sources as some may be coming from outside	Used to make polystyrene, Solvents, Vinyl and PVC, Nylon carpets, furnishings
Toluene (TOL)	Exposure to toluene can cause eye and nose irritation, tiredness, confusion, euphoria, dizziness, headache, dilated pupils, tears, anxiety, muscle fatigue, insomnia, nerve damage, inflammation of the skin and liver and kidney damage	Solvent in paints, lacquers, glues, thinners A higher emitter following recent decoration work
Xylenes (XYL)	Irritate the eyes, nose, skin and throat. Xylene can also cause headaches, dizziness, confusion, loss of muscle coordination and in high doses, death	A solvent and thinning agent, used to make polyester fibre and used in sealers in concrete and waterproofing
Chloroform (CFORM)	Symptoms include excitement and nausea followed by dizziness and drowsiness. More severe exposures to chloroform may cause heart problems, fitting, unconsciousness and in some cases death	Used as a solvent in paper and board materials, floor polishes, resins adhesives and rubber
Chlorinated solvents	Can readily volatilise, do not break down easily in the environment and a small amount can contaminate a large area. Because of this, chlorinated solvents can cause significant environmental problems, dizziness, fatigue, headaches and/or skin rashes. Long-term side effects may include chronic skin problems, and/or damage to the nervous system, kidneys or liver. Some chlorinated solvents are also known to cause cancer	Used in thermoplastics and PVC

Table 2.2 (Continued)

List of chemical contaminants widely used in building materials	*Potential health impacts*	*Use in building materials*
Methylene chloride (MECL) Dichhloromethane	Can harm the eyes, skin, liver, and heart. Exposure can cause drowsiness, dizziness, numbness and tingling limbs, and nausea. It may cause cancer Proposals exist for it to be banned	Paint strippers banned for domestic use however it is relatively easy to obtain and is used to create decaffeinated coffee!
Perchloroethylene (PERC) Tetrachloroethylene	Short-term exposure to high levels of perchloroethylene causes dizziness, sleepiness, confusion, headache, and eye, nose and throat irritation. Exposed people may have trouble speaking or walking or may lose consciousness. Exposure to intensely high levels for a short time can be deadly	Dry cleaning solvent Paint and spot removers, water repellents, wood cleaners, glues
Acetone 2-propanone or dimethyl ketone	Can cause irritation of skin	Used in production of acrylic plastics, polycarbonates and epoxy resins
Methyl ethyl ketone (MEK) Butanone	Inflammation of mouth and stomach, lung damage, skin irritation	Solvent in paints, plastics and adhesives
Cyclohexanone (CHEX)	Solvent	Solvent can affect the nervous system
Ethyl acetate (EtOAc)		Component in varnishes ad thinners, epoxies, acrylics, urethanes and vinyl
Lifestyle, cleaning products Fragrances, preservatives, UV filters & disinfectants		
Ethyl alcohol (EOH) Ethanol Butyl acetate (BuAc) nicotine (NIC) 10 0 Tonalide (AHTN) Galaxolide (HHCB) Musk ketone (MK) Musk xylene (MX) Methyl Paraben (MePa) Benzophenonee (BP) Benzophenone-3 (BP-3) Triclosan (TCS) Cleaning		Mainly used in cosmetics and perfumes
Insulation		
Styrenes A derivative of benzene	"Styrene sickness" headache, nausea, vomiting, weakness, fatigue, dizziness ataxia, irritation of the nose and throat, increased nasal secretion, wheezing, coughing, pulmonary oedema, cardiac arrhythmias and coma	to make latex, synthetic rubber, and polystyrene resins. Insulation Styrene can occur naturally in some plants such as strawberries so some companies state that it is therefore safe

(*Continued*)

Table 2.2 (Continued)

List of chemical contaminants widely used in building materials	*Potential health impacts*	*Use in building materials*
Butadiene Isoprene 2-methyl-1,3-butadiene	VOC lethargy, headache, drowsiness, fatigue, vertigo, ataxia, unconsciousness, coma, and respiratory depression and death. Highly flammable	Used in rubber. Styrene butadiene rubber, water proofing with cement
Acrylonitrile	Headaches, dizziness, confusion, vomiting	Mainly used in clothing, carpets and upholstery
Diethylene glycol (DEG)	Cancer risk. It is also explosive and highly flammable. Is absorbed into the nervous system	A solvent used in resins Polyurethane insulation solvent
Oxirane Ethylene Oxide Ethylene glycol	Classified as highly toxic. respiratory problems	Spray foam insulation and foam fire extinguishers also a pesticide
Propylene Oxide		Used in spray foam insulation
Polyurethane (TPU)	Powerful irritant to the mucous membranes of the eyes, gastrointestinal tract, and respiratory system. Frequent exposure leas to sensitisation	Insulation, fillers and sealants, adhesives. Made from Isocyanates combined with oil (polyols) with a range of metal catalysts
Isocyanates MDI and TDI	Chest tightening, skin bronchial irritation, dermatitis cancer, asthma mainly a problem for workers spraying isocyanates such as spray foam	MDI dark brown liquid or in white or yellow flakes. TDI colourless to yellow liquid. Extremely volatile Insulation component and widely used in sealants and adhesives and paints
Methyl Isocyanate	Isocyanates are made from Amines treated with Phosgene. Highly toxic gas. Bhopal Disaster which killed at least 3,800 people	Widely used in insulation, adhesives and sealants
Cyanoacrylics Adhesive	Asthma and nasal irritation	Used in adhesives
Acetate Ethylene	Irritation of various organs including nervous symptoms	
Polyvinyl Acetate and Vinyl Chloride	A danger in terms of its impact on the wider environment due to dioxins Released in large quantities in the Ohio train crash. Causes irritation with chronic exposure cancer and neurological problems	Cladding, wiring, windows and doors, flooring
Isobutylene Polyisobutylene	Irritant, dizziness and fatigue	Coatings, sealants, mastics, glues
Butyl Acrylate	Can cause skin allergy, dizziness and vomiting. Highly flammable	Used in water-based paints adhesives and various finishing materials, polymers and resins

Table 2.2 (Continued)

List of chemical contaminants widely used in building materials	*Potential health impacts*	*Use in building materials*
Napthalene (Moth balls)	Highly toxic. Acute effects even with short exposure. Gastrointestinal problems haemolytic anaemia, damage to the liver, and neurological damage. Probable carcinogen, no safe level of exposure	Widespread use as plasticiser, pesticide, surfactant use in epoxy resin adhesives
Microsynthetic fibres	Inflammation, cancer and reproductive problems and DNA damage Major concern in terms of widespread pollution	Used in concrete, polyesters, nylon, polypropylene, mats, upholstery, cleaning products
PFAS per- and polyfluoroalkyl substances	Cancer, liver damage, decreased fertility, and increased risk of asthma and thyroid disease	Surface sealers for stone, tile, grout, caulks and concrete and increasingly used in other products such as carpets and rugs. Plastic coatings paints. Roofing materials, weatherproofing. Wood flooring adhesives, windows and glass, ceramic surfaces fabrics wires and cables. PTFE plumber's tape, sealing tapes for air tightness and membranes, solar panels, artificial grass seismic dampers in buildings prone to earthquakes
Fungicides (pesticides in building materials) azoxystrobin, pyraclostrobin, fluoxastrobin and trifloxystrobin captan, folpet, dithiocarbamates, pentachlorophenol, and mercurials. Stobilurin thiabendazole	stinging eyes, rashes, blisters, blindness, nausea, dizziness, diarrhoea and death. chronic effects are cancers, birth defects, reproductive harm, immunotoxicity, neurological and developmental toxicity, and disruption of the endocrine system. House dust can be heavily contaminated with fungicides from wall boards	Used widely now in wall boards and "plasterboard" Microban coatings Sealants, timber treatments
Silver Nanoparticles	May cause death of brain cells and tissue with long exposure Can easily penetrate human tissues Skin allergen	Used as an anti-fungal in paints and coatings
Silica (found in house dust tests)	Fine dust can easily het into lungs COPD kidney disease, auto immune diseases, Silicosis which is debilitating and can be fatal. It is still common to see building workers creating dust while not wearing protective masks	Major constituent of masonry materials, only a problem with dust from cutting, drilling, grinding and polishing

(*Continued*)

Table 2.2 (Continued)

List of chemical contaminants widely used in building materials	*Potential health impacts*	*Use in building materials*
Asbestos (still widely found especially in schools and the Houses of Parliament)	Lung and other cancers, mesothelioma	The use of asbestos in buildings and materials was not fully banned until 1999. It is still widely prevalent in many older buildings and is still used today in the manufacture of roofing compounds gaskets and friction products. Asbestos is still widely used in composite product manufacture around the world which may be imported into the UK
Lead	Headaches, tiredness, irritability, anaemia, stomach pains, kidney, nerve and brain damage, and even possibly cancer	Prevalent in older houses in paint but widely used in other building materials and still found in water pipes
Other heavy metals, as additives and stabilisers Mercury Arsenic, Antimony, Cadmium, Chromium, Copper, Cobalt, Zinc	Gastrointestinal kidney, skin disorders vascular damage	Can be used in many composite materials though rarely declared
Carbon Black	Lung disease, irritation	Cement, renders, plasters and screeds
Graphene	Lung injury and suspected risks from nanoparticles	Used in paints and insulation
Benzene & Propane	Cancer	Asphalt, glues and paints
Toluene	Eye and nose irritation, tiredness, confusion, euphoria, dizziness, headache, dilated pupils, tears, anxiety, muscle fatigue, insomnia, nerve damage, inflammation of the skin, and liver and kidney damage	Paints and lacquers glues and adhesives
Xylenes	Irritate the eyes, nose, skin, and throat. Xylene can also cause headaches, dizziness, confusion, loss of muscle coordination, and in high doses, death.	Solvent. Used to make polyester fibre, dies, paints
Plasticisers, Pthalates and Bisphenol A	Reproductive and developmental disorders including infertility and early puberty	Adhesives and sealants and PVC
Solvents including xylene, toluene listed above	High airborne concentrations of some solvents can cause unconsciousness and death. Exposure to lower levels of solvents can lead to short-term effects including irritation of the eyes, lungs and skin, headaches, nausea, dizziness or light-headedness, kidney damage	White Spirit, varnishes, cleaning product adhesives
Mould	Asthma, allergies, lung disease	See Chapter 5

Table 2.2 (Continued)

List of chemical contaminants widely used in building materials	*Potential health impacts*	*Use in building materials*
Biocides and Pesticides • Benzimidazoles • Carbamates • Disinfectants • Triazines • Triazole • Mecoprop. • 2-Methyl-4-chlorophenoxyacetic Acid. • Carbendazim • Terbutryne • Ammonium compounds • Pyrethroids • Pyrithione	Effects on pregnant women, skin respiratory and nervous system Carcinogenic and endocrine disrupting Cardiovascular disease and stroke. It may take 20 years for problems to occur *Application, release, Ecotoxicological assessment of biocide in building materials and its soil microbial response* Ecotoxicology and Environmental Safety 224 (2021) 112707	Biocides have been shown to leech into the sewage system and to soils. They are used in building materials in an attempt to inhibit mould, mildew and algae such as in plasterboard and gypsum plaster Many building materials are now using biocides as a mould inhibitor
Surfactants PVA polyoxyethylene glycol monomethyl ether (POE).	Detergents are generally not too dangerous but much stronger chemicals are used in building materials surfactants today, but it was not possible to find health risk literature. Some surfactants are used in medical treatment, but they can be dangerous for young children	Used in cleaning, surfactants reduce surface tension increasing its wetting properties. Used in lightweight concrete, mortar, plasticisers paints. (Building workers used to put washing up liquid into cement mixes)
Driers Metals and acids Cobalt, Manganese, Cerium, Zirconium 2-ethylhexanoate (octoate, C8) l Neodecanoate (C10) l Isononanoate (C9 Metal carboxylates are more commonly referred to as driers or siccatives	Cobalt is listed as hazardous with lung, allergy and hearth problems from exposure Lead is no longer used in paints but there are many chemicals used. There are cancer risks for painters working with materials every day	Used to speed drying in plastic (alkyd) paints
Flame retardants	The aim of flame retardants is to raise the temperature at which a material ignites, reduce the rate at which materials burn and minimise spreads of flame. Given the high number of fires and the speed at which fires spread it is unknown whether flame retardants are effective given the similar falls in fire deaths across countries, with or without, open flame tests	

(*Continued*)

Table 2.2 (Continued)

List of chemical contaminants widely used in building materials	*Potential health impacts*	*Use in building materials*
Polybrominated BDE 47e 10 0 2,3-dibromo-1-propanol (23DB1P) 2,2-bisbromomethyl-1,3-propanediol – 1,3-dichloro-2-propanolg (13DC2P) tris(1-chloro-2-propyl) phosphate (TCIPP) tris(1,3-dichloroisopropyl) phosphate (TDCIPP) tris(2-chloroethyl) phosphate (TCEP) tris(2-butoxyethyl) phosphate (TBOEP) triphenyl phosphate (TPHP) DBDPE which replaced decaBDE This is not a complete list. Many more could be included.	Flame retardants cause neurological damage, hormone disruption and cancer. They can bio accumulate in the bodies and most people now have contaminants from flame retardants in their blood. Environments International Volume 173 March 2023 107782. A new consensus on reconciling fire safety with environmental and health impacts of chemical flame retardants	Flame retardants are used in all synthetic insulation materials. Details are often not disclosed in product data sheets. Mineral fibre insulations are described as fire safe, but they can dissolve in fires
SVOCs Semi volatile organic compounds are a subgroup of VOCs that have a higher molecular weight and a higher boiling point temperature. They should not be confused with VOCs Perfluoroalkyl acids (PFAS) (see above) Polybrominated flame-retardants (see above under flame retardants)		
Phthalates/plastics chemicals diethyl phthalate (DEP) di-n-butyl phthalate (DBP) butylbenzyl phthalate (BBP) bis(2-ethylhexyl) adipate (DEHA) bis(2-ethylhexyl) phthalate (DEHP) dicyclohexyl phthalate (DCHP) diisononyl phthalate (DINP) 10	Damage to liver, kidneys, lungs and reproductive system	Adhesives and sealants, used in PVC

Table 2.2 (Continued)

List of chemical contaminants widely used in building materials	*Potential health impacts*	*Use in building materials*
Polychlorinated biphenyls (PCBs)	PCBs are accepted as carcinogenic with damaging effects on the immune, reproductive, neurological and endocrine system can cause skin rashes as well as being regarded as a persistent organic pollutant in the wider environment. Banned by the Stockholm conference in 2001 but production was stopped in USA in 1979. PCBs will still exist in buildings built up to 1970	Electrical equipment, caulks and mastics, tiles carpets, adhesives and paints
Polyaromatic hydrocarbons (PAHs)	PAHs result from burning coal, oil, gas, wood, rubbish, and tobacco and ambient air. PAHs can bind to or form small particles in the air. High heat when cooking meat and other foods will form PAHs. This has led to the suggestion that certain types of cooking create a particulate health risk in the home. PAHs can be carcinogenic but can also have pulmonary and gastrointestinal effects	Not used in building materials
Flooring		
Hexamethylene diisocyanate (HDI)	Acute (short-term) exposure to high concentrations of hexamethylene diisocyanate in humans can cause pulmonary edema, coughing, and shortness of breath	Acrylic flooring adhesive, wood, cork, vinyl, possibly in carpet. Also, used in polyurethane paints and coatings
Ethyl Carbamate	Cancer	Solvent wood, cork rubber
Formaldehyde	Cancer	Preservative, wood, cork bamboo, fluid applied
Methylene bisphenol diisocyanate (MDI	Skin and throat irritant, lack of research into long term health effects	Flooring
Ethylenelmine Azirdine	Cancer	Flooring
N-Methyl-2-pyrrolidone (NMP)		Flooring
Ethylene glycol monbutyl ether (EGE)	Cancer	Cork flooring solvent
Butyl benzyl phthalate BBP	Cancer	Flooring plasticiser
Vinyl Chloride	Cancer	Flooring sheet and tiles widely used in hospitals
Vinyl Acetate	Cancer	Flooring performance enhancer
1,6-Hexanediamine	Skin irritation, possible burns, respiratory and digestive tract irritation.	Carpet

(*Continued*)

Table 2.2 (Continued)

List of chemical contaminants widely used in building materials	*Potential health impacts*	*Use in building materials*
Maleic anhydride	Asthma and allergies, coughing	Flooring
Ethanol	Breathing problems, low blood pressure liver damage	Flooring pigment solvent
Butylated hydroxytoluene (BHT)	Cancer	Carpet and flooring antioxidant
1,3-Butadiene	Cancer	Flooring
Melamine	Cancer	
Urea formaldehyde	Toxic, releases free formaldehyde possible carcinogen	Scavenger wood and cork flooring and furniture. Was used in insulation in cavity walls
Aminoethylpiperazine	Irritant, headache, nausea and vomiting	Fluid applied flooring epoxy
Ethanolamine	Severe irritant and can cause allergies	Cork flooring surface active agent
Methyl ethyl Ketoxime (MEKO)	Cancer	Wooden flooring blocking agent in coatings
Benzo[g,h,i] perylene	Suspected carcinogen but limited data available	Carpet
Diethylene glycol monoethyl ether	Renal, hepatic and testicular toxicity	Water born flooring finish
Texanol 2,2,4-Trimethyl-1,3-pentanediol monoisobutyrate	Possible carcinogen	Wood flooring solvent
Ethylbenzene	Cancer	Flooring solvent
Concrete, cementitious screeds and renders	Cement and sand have been used for floor screeds for many years. In recent years, many different chemicals have been added to proprietary screed mixtures for a variety of reasons, ease of application, self-levelling, quick drying, etc. Many additives are classified as bio-accumulative and toxic, subject of very high concern under REACH, reproductive toxicity	
Carbon Black	Skin irritation, eye damage lung disease	Used in screeds as well as some insulation products, paints sealants and adhesives, fibres and rubber
Magnetite	Very little information available on health risks	Used to make steel. Used in composites to make screeds and plasters. Can be used in walls to resist x-rays

Table 2.2 (Continued)

List of chemical contaminants widely used in building materials	*Potential health impacts*	*Use in building materials*
Blast furnace slag and Fly Ash	Risk is mainly from dust which can cause irritation	Used to produce s-called low carbon cements and concrete
Screed additives can contain 5-chloro-2-methyl-2H-isothiazol-3-one 2-methyl-2H-isothiazol-3-one 1,2-benzisothiazol-3(2H)-one 2,2,4-trimethyl-1,3-pentanediol monoisobutyrate Styrene	It has been difficult to find health effect data but possible acute toxicity	Screed additive
Hydrogen sulphide and sulphur dioxide	Hydrogen sulphide is a highly toxic and poisonous gas can be released from slag cement	Used in cements and plaster and plasterboard materials made from power station waste, the sulphur is meant to be removed through "desulphurisation" such as from Drax Power Station **(120)**, **(121)**

Firefighters three times more likely to die from certain types of cancer – study

Cancer death rates are 1.6 times higher than general population probably due to exposure to toxic fumes. **(113)**

Cancer among fire fighters is a growing problem as various studies have followed the health of fire fighters. A report by the Fire Brigades Union claims that up to a dozen firefighters who responded to the 2017 Grenfell Tower fire have been diagnosed with digestive cancers or leukaemia. Media reports have claimed that Firefighters three times more likely to die from certain types of cancer (some say 6 times) and that cancer death rates among fire fighters are 1.6 times higher than general population probably due to their greater exposure to toxic chemicals. **(114)**, **(115)**, **(Op Cit 113)**

References

(1) Consumer Protection, Environmental Protection Health and Safety, The REACH (Amendment) Regulations 2023. (n.d). Available at:www.legislation.gov.uk/ukdsi/2023/9780348247329/pdfs/ukdsi_9780348247329_en.pdf [Accessed 15 Sep. 2023].

(2) Your Europe. (n.d). Registering chemicals in the EU (REACH regulation). Available at: https://europa.eu/youreurope/business/product-requirements/chemicals/registering-chemicals-reach/index_en.htm [Accessed 15 Sep. 2023].

(3) 'Safety Data Sheet (SDS) in Accordance With Commission Regulation (EU) No 453/2010 (Reach).' (n.d) [pdf]. Available at: www.generalinsulation.com/wp-content/uploads/2015/09/Rockwool-Mineral-Wool-SDS.pdf [Accessed 15 Sep. 2023].

(4) Wenger, C. et al. (2014). Disarming reach – an insight to lobbying in the European Union. *Academia.edu*. Available at: www.academia.edu/8515935/Disarming_REACH_An_Insight_to_Lobbying_in_the_European_Union [Accessed 13 Sep. 2023].

(5) Lewitin, J. (2022). 'How does vinyl flooring impact the environment?' *The Spruce*. Available at: https://www.thespruce.com/environmental-impact-of-vinyl-flooring-1314956 [Accessed 13 Sep. 2023].

(6) Kraus, M. and Senitkova, I.J. (2020). 'VOCs emission simulation of common flooring materials', *IOP Conference Series: Materials Science and Engineering*, 960(4). doi:10.1088/1757-899x/960/4/042093.

(7) Adamová, T., Hradecký, J., and Pánek, M. (2020). 'Volatile Organic Compounds (VOCs) from wood and wood-based panels: Methods for evaluation, potential health risks, and mitigation', *Polymers*, 12(10), 2289. doi:10.3390/polym12102289.

(8) Hodgson, A. T., Wooley, J. D., and Daisey, J. M. (1993). 'Emissions of volatile organic compounds from new carpets measured in a large-scale environmental chamber', *Air & Waste: Journal of the Air & Waste Management Association*, 43(3), 316–324. Available at: https://doi.org/10.1080/1073161x.1993.10467136

(9) Mulrenan, R. (2019). 'What's In Your Carpet?' *Wen*. Available at: www.wen.org.uk/2019/01/03/whats-in-your-carpet/ [Accessed 15 Sep. 2023].

(10) Healthy Building Network. (2017). Eliminating Toxics in Carpet: Lessons for the Future of Recycling. Available at: https://healthybuilding.net/blog/227-eliminating-toxics-in-carpet-lessons-for-the-future-of-recycling [Accessed 15 Sep. 2023].

(11) Schieweck, A. and Bock, M.-C. (2017). 'Emissions from low-VOC and zero-VOC paints – Valuable alternatives to conventional formulations also for use in sensitive environments?' *Building and Environment*, 85, 243–252. Available at: https://doi.org/10.1016/j.buildenv.2014.12.001 [Accessed 15 Sep. 2023].

(12) Allergy Free. (n.d). Paint Allergy Symptoms and Management – Allergyfree. Available at: www.allergyfree.co.in/en-gb/know-your-allergy/indoor-allergies/paint#:~:text=Inhalation%20of%20paint%20fumes%20can [Accessed 15 Sep. 2023].

(13) Ines. (2023). 'Can You Be Allergic To Acrylic Paint? | Sustain The Art,' *Sustain The Art*. Available at:https://sustaintheart.com/can-you-be-allergic-to-acrylic-paint/?utm_content=cmp-true[Accessed 15 Sep. 2023].

(14) ICI Paints Akzo Nobel. (n.d). 'Safety Data Sheet| Durable Acrylic Eggshell' [pdf]. Available at: https://prdakzodecodocumentssa.blob.core.windows.net/public/clp_msds/am/uk/en/am_gb_en_durable_acrylic_eggshell.pdf [Accessed 15 Sep. 2023].

(15) The Essential Chemical Industry. (2013). Paints. Available at: www.essentialchemicalindustry.org/materials-and-applications/paints.html#:~:text=Decorative%20gloss%20paints%20typically%20contain,(linseed%20or%20soybean%20oil). [Accessed 15 Sep. 2023].

(16) New Jersey Department of Health. (2016). Right to Know Hazardous Substance Fact Sheet. Available at: https://nj.gov/health/eoh/rtkweb/documents/fs/1861.pdf [Accessed 15 Sep. 2023].

(17) Swoboda, F. (1993). 'U.S Rethinks Calling Fiber Glass Possible Carcinogen', *Washington Post*. Available at: www.washingtonpost.com/archive/business/1993/09/10/us-rethinks-calling-fiber-glass-possible-carcinogen/5fa75b5c-d921-495c-abc1-34ea050b3933/ [Accessed 15 Sep. 2023].

(18) Infante, P. F., Schuman, L. D., and Huff, J. (1996). 'Fibrous glass insulation and cancer: Response and rebuttal', *American Journal of Industrial Medicine*, 30(1), 113–120. https://doi.org/10.1002/(sici)1097-0274(199607)30:1<113: aid-ajim21>3.0.co;2-# [Accessed 15 Sep. 2023].

(19) Campopiano, A. et al (2012). 'Glass fibre exposure assessment during ceiling installation by European Standard EN 689: study of airborne fibre distribution', *Atmospheric Pollution Research*, 3(2), 192–198. doi:https://doi.org/10.5094/apr.2012.020

(20) Knauf Insulation. (n.d). Earthwool Insulation Earthwool Product Changes. Available at: www.knaufinsulation.co.uk/earthwool-product-updates [Accessed 15 Sep. 2023].

(21) Sullivan, P. (2019). 'Insulation plant divides community in West Virginia', *Washington Post*. Available at: www.washingtonpost.com/local/virginia-politics/insulation-plant-divides-community-in-west-virginia/2019/06/09/d0ea34a8-7d68-11e9-8ede-f4abf521ef17_story.html [Accessed 15 Sep. 2023].
(22) West Virginia Department of Environmental Protection. (2017). 'Division of Air Quality' [pdf]. Available at: https://dep.wv.gov/daq/Documents/November%202017%20Permits%20and%20Evals/777-00086_PERM_13-3376.pdf [Accessed 15 Sep. 2023].
(23) Neuhaus, T., Oppl, R., and Clausen, A. (2008). 'Formaldehyde emissions from mineral wool in building constructions into indoor air', *Indoor Air.* Available at: www.eurofins.com/media/2266/formaldehyde_emissions_from_mineral_wool_in_building_constructions_1086.pdf [Accessed 15 Sep. 2023].
(24) Bennett, T.M et al. (2022). 'Low formaldehyde binders for mineral wool insulation: A review.' *Global Challenges*, 6(4). doi: https://doi.org/10.1002/gch2.202100110.
(25) Rockwool. (n.d). Frequently Asked Questions. Available at: www.rockwool.com/north-america/advice-and-inspiration/faq/ [Accessed 15 Sep. 2023].
(26) Bridgend County Borough Council. (n.d). Agenda Item 8 [pdf]. Available at: https://democratic.bridgend.gov.uk/documents/s17132/Appendix%201%20-%20Rockwool%20DC%20Committee%20Report%20P18520FUL.pdf [Accessed 15 Sep. 2023].
(27) European Certification Board for Mineral Wool Products. (n.d). Who's who? Available at: www.euceb.org/index.php?page=who-s-who [Accessed 15 Sep. 2023].
(28) Mirata, S. et al (2022). 'The acute toxicity of mineral fibres: A Systematic in vitro study using different THP-1 macrophage phenotypes.' *International Journal of Molecular Sciences*, 23(5). Available at: https://doi.org/10.3390/ijms23052840 [Accessed 15 Sep. 2023].
(29) Yliniemi, J. et al (2021). 'Characterization of mineral wool waste chemical composition, organic resin content and fibre dimensions: Aspects for valorisation.' *Waste Management (New York, N.Y.)*, 131, 323–330. Available at: https://doi.org/10.1016/j.wasman.2021.06.022 [Accessed 15 Sep. 2023].
(30) Lucas, A. (2014). 'Styrene and polystyrene foam 101.' *Toxic-Free Future.* Available at: toxicfreefuture.org/blog/styrene-and-styrofoam-101-2/# [Accessed 15 Sep. 2023].
(31) Jablite Intelligent Insulation. (2015). 'Jablite LIMITED Environmental, Health and Safety Data Sheet Expanded Polystyrene' [pdf]. Available at: www.dryvit.co.uk/media/14958/duk_sds_eps_all.pdf [Accessed 15 Sep. 2023].
(32) New Jersey Department of Health. (2009). 'Right to Know Hazardous Substance Fact Sheet| Pentane' [pdf]. Available at: https://nj.gov/health/eoh/rtkweb/documents/fs/1476.pdf [Accessed 15 Sep. 2023].
(33) World Centric. (n.d). Impacts and Risks of Polystyrene. Available at: www.worldcentric.com/journal/impacts-and-risks-of-styrofoam [Accessed 15 Sep. 2023].
(34) Hwang, J. et al (2020). 'Potential toxicity of polystyrene microplastic particles.' *Scientific Reports*, 10(1), 1–12. doi:https://doi.org/10.1038/s41598-020-64464-9.
(35) Booth, R. (2020) 'Makers of Grenfell cladding abused testing regimes, inquiry told.' *The Guardian*. Available at: www.theguardian.com/uk-news/2020/nov/05/makers-of-grenfell-cladding-abused-testing-regimes-inquiry-told [Accessed 15 Sep. 2023].
(36) Federal Trade Commission. (2006). FTC Stops Allegedly False Claims About Insulation Performance. Available at: www.ftc.gov/news-events/news/press-releases/2006/10/ftc-stops-allegedly-false-claims-about-insulation-performance [Accessed 15 Sep. 2023].
(37) O'Loughlin, A. (2017). 'Building company sues firms over allegedly defective insulation boards which 'caused the floor to sink'. *Irish Examiner.* Available at: www.irishexaminer.com/news/arid-30817093.html [Accessed 15 Sep. 2023].
(38) Carolan, M. (2017). 'Kelland Homes sues over allegedly defective insulation boards.' *The Irish Times*. Available at: www.irishtimes.com/business/construction/kelland-homes-sues-over-allegedly-defective-insulation-boards-1.3315217 [Accessed 15 Sep. 2023].

(39) Brennan, J. (2017). 'Builders face insulation 'headache' amid key chemical shortage.' *The Irish Times.* Available at: www.irishtimes.com/business/construction/builders-face-insulation-headache-amid-key-chemical-shortage-1.3075944 [Accessed 15 Sep. 2023].

(40) Chemanager International. (2015). Covestro to Close Tarragona MDI Plant. Available at: www.chemanager-online.com/en/news/covestro-close-tarragona-mdi-plant [Accessed 15 Sep. 2023].

(41) Herrero, A. (2019). Chemical Industry in Flix, Tarragona, Spain. Available at: https://ejatlas.org/conflict/chemical-industry-in-flix-tarragona-spain [Accessed 15 Sep. 2023].

(42) ERCROS. (n.d). Estimated impact to ERCROS of banning the use of mercury technology. Available at: www.ercros.es/index.php?option=com_content&view=article&id=1183:estimated-impact-to-ercros-of-banning-the-use-of-mercury-technology-and-covestro-plant-shutdown-from-december-2017&catid=44:relevant-events&Itemid=101&lang=en [Accessed 15 Sep. 2023].

(43) PU Europe Factsheet. (2015). Indoor air quality and polyurethane insulation [pdf]. Available at: www.pu-europe.eu/wp-content/uploads/2023/06/Factsheet_18_Indoor_air_quality_and_polyurethane_insulation.pdf [Accessed 15 Sep. 2023].

(44) Lockey, J. E. et al (2015). 'Isocyanates and human health: multi-stakeholder information needs and research priorities.' *Journal of Occupational and Environmental Medicine*, 57(1), 44–51. Available at: https://doi.org/10.1097/JOM.0000000000000278 [Accessed 15 Sep. 2023].

(45) Broughton, E. (2005). The Bhopal disaster and its aftermath: A review. *Environmental Health: A Global Access Science Source*, 4(1), 6. Available at: https://doi.org/10.1186/1476-069X-4-6 [Accessed 15 Sep. 2023].

(46) U.S Environmental Protection Agency. (1985). Locating and Estimating Air Emissions from Sources of Phosgene [pdf]. Available at: www.epa.gov/sites/default/files/2020-11/documents/phosgene.pdf [Accessed 15 Sep. 2023].

(47) Greene, M. G. (2013) 'Spray foam insulation can make some homes unliveable.' *CBC News.* Available at: www.cbc.ca/news/spray-foam-insulation-can-make-some-homes-unlivable-1.222428 [Accessed 15 Sep. 2023].

(48) Blitzer, R. (2016) 'Whistle-blowers sue chemical companies over alleged failure to disclose health risks.' *Law & Crime.* Available at: http://lawnewz.com/high-profile/whistleblowers-sue-chemical-companies-over-alleged-failure-to-disclose-health-risks/ [Accessed 15 Sep. 2023].

(49) Healthy Indoors. (2016). Lawsuit Against Four Companies Over Spray Foam. Available at: https://healthyindoors.com/2016/09/lawsuit-against-four-chemical-companies-over-spray-foam-insulation/ [Accessed 15 Sep. 2023].

(50) UK Parliament. (2023). Spray Foam Insulation and Mortgages. Available at: https://commonslibrary.parliament.uk/spray-foam-insulation-and-mortgages/ [Accessed 15 Sep. 2023].

(51) Holladay, M. (2011). 'Three Massachusetts Home Fires Linked to Spray-Foam Installation.' *Green Building News.* Available at: www.greenbuildingadvisor.com/blogs/dept/green-building-news/three-massachusetts-home-fires-linked-spray-foam-installation [Accessed 15 Sep. 2023].

(52) Wirts, M., Salthammer, T. (2002). 'Emission of Isocyanates from PUR-Adhesives. Time Course of Emission From Curing PUR-Adhesive.' [pdf] *Indoor Air.* Available at: www.irbnet.de/daten/iconda/CIB7080.pdf [Accessed 15 Sep. 2023].

(53) Wirts, M. (2003). 'Time course of isocyanate emission from curing polyurethane adhesives.' *Atmospheric Environment*, 37(39–40), 5467–5475.

(54) Keough, C. (1980). 'Urea-formaldehyde foam: A dangerous situation.' *DOE Office of Scientific and Technical Information.* Available at: www.science.gov/topicpages/u/urea-formaldehyde+foam+insulation [Accessed 15 Sep. 2023].

(55) (2016). 'Revealed: £1 million cost of Flash Ley School toxic gas closure.' *Express & Star.* Available at: www.expressandstar.com/news/education/2016/07/21/revealed-1-million-cost-of-flash-ley-school-toxic-gas-closure/ [Accessed 15 Sep. 2023].

(56) Aquaresins Technologies. (n.d). *Benefil RG22 hardfoam.* Available at: https://aquaresinstechnologies.com/products/rg22/ [Accessed 15 Sep. 2023].

(57) (2015). 'Parents urged to attend public meetings about school toxic gas after high levels of formaldehyde makes pupils ill.' *ITV News*. Available at: www.itv.com/news/central/2015-10-09/public-meeting-about-toxic-gas-scare-at-stafford-school/ [Accessed 15 Sep. 2023]

(58) (2015). 'Schools closed and pupils ill after toxic gas detected.' *ITV News*. Available at: www.itv.com/news/central/2015-10-08/school-closed-and-pupils-ill-after-toxic-gas-leak/ [Accessed 15 Sep. 2023].

(59) CAPITA. (2022). Education Services. Available at: www.capita.com/expertise/education/education-services [Accessed 15 Sep. 2023].

(60) ENTRUST (n.d). Design and Property Management. Available at: www.entrust-ed.co.uk/services/design-and-property-management [Accessed 15 Sep. 2023].

(61) From Public Health England FOI letter to the author 20.4.18 (ref 26/03/Ih/997).

(62) National Cancer Institute. (2011). Formaldehyde and Cancer Risk. Available at: www.cancer.gov/about-cancer/causes-prevention/risk/substances/formaldehyde/formaldehyde-fact-sheet [Accessed 15 Sep. 2023].

(63) Knufken, D. (2009). 'Knauf Drywall Fiasco 'Biggest Home Defect Case in US History." *Business Pundit.* Available at: www.businesspundit.com/knauf-drywall-fiasco-biggest-home-defect-case-in-us-history/ [Accessed 15 Sep. 2023].

(64) Cushman, T. (2012). 'Knauf Settles with Chinese Drywall Homeowners.' *The Journal of Light Construction.* Available at: www.jlconline.com/how-to/interiors/knauf-settles-with-chinese-drywall-homeowners_o [Accessed 15 Sep. 2023]

(65) United States District Court For The Eastern District of Louisiana, (n.d) 'Sean and Beth Payton v. KNAUF GIPS KG; KNAUF PLASTERBOARD (TIANJIN) CO., LTD.; KNAUF PLASTERBOARD (WUHU), CO., LTD.; KNAUF PLASTERBOARD (DONGGUAN) CO., LTD' [pdf]. Available at: http://cdnassets.hw.net/e4/a7/0fa72bb64c2d912045498c6935cb/peyton-v-knauf.pdf [Accessed 15 Sep. 2023].

(66) Sapien, J. (2010). 'Is Chinese drywall making habitat for humanity's houses uninhabitable?' *ProPublica.* Available at: www.propublica.org/article/is-chinese-drywall-making-habitat-for-humanitys-houses-uninhabitable [Accessed 15 Sep. 2023]

(67) Fava, A. (2017). £60k fine for Plasterboard Recycling Form Following Inspections. Available at: www.shponline.co.uk/common-workplace-hazards/60k-fine-plasterboard-recycling-firm-following-inspections/ [Accessed 15 Sep. 2023]

(68) (2014). 'Gypsum Board: Are Our Walls Leaching Toxins.' *Building Green.* Available at: https://www.buildinggreen.com/blog/gypsum-board-are-our-walls-leaching-toxins [Accessed 15 Sep. 2023].

(69) (n.d). 'Environment Agency permitting decisions.' [pdf] Available at: https://assets.publishing.service.gov.uk/government/uploads/system/uploads/attachment_data/file/339414/Decision_Document.pdf [Accessed 15 Sep. 2023].

(70) Murtoniemi, T., Nevalainen, A., and Hirvonen, M. R. (2003). 'Effect of plasterboard composition on Stachybotrys chartarum growth and biological activity of spores.' *Applied and Environmental Microbiology*, 69(7), 3751–3757. Available at: https://doi.org/10.1128/AEM.69.7.3751-3757.2003 [Accessed 15 Sep. 2023].

(71) (2018). 'Gaelic football club sinkholes 'caused by mine work.' *BBC News Europe.* Available at: www.bbc.com/news/world-europe-45665611 [Accessed 15 Sep. 2023].

(72) Elvins, J. (n.d). 'Natural gypsum bounces back.' Available at: www.mineralandwasteplanning.co.uk/natural-gypsum-bounces-back/energy-minerals/article/1435302 [Accessed 15 Sep. 2023]

(73) Bhat, A. (2012). 'Gypsum Extraction in Kashmir Impacts Health, Environment.' *Global Press Journal.* Available at: https://globalpressjournal.com/asia/indian-administered_kashmir/gypsum-extraction-in-kashmir-impacts-health-environment [Accessed 15 Sep. 2023].

(74) LANXESS. (n.d). Dry-wall and Finishing. Available at: https://lanxess.com/en/Products-and-Brands/Industries/Microbial-Control/Industries-Overview/Construction/Dry-wall-and-finishing [Accessed 15 Sep. 2023].

(75) LANXESS (n.d). Sporgard. Available at: https://lanxess.com/en/Products-and-Brands/Products/s/SPORGARD--WB [Accessed 15 Sep. 2023].

(76) Weigl, M. et al (2014). 'VOC emissions from building materials: results from lab and model room trials.' *International Wood Products Journal*, 5(3), 136–138. Available at: https://doi.org/10.1179/2042645314y.0000000072 [Accessed 15 Sep. 2023].

(77) Liddell, H., Gilbert, J., and Halliday, S. (2008). 'Design and Detailing for Toxic Chemical Reduction in Buildings.' [pdf] *SEDA,* no. 3. Available at: https://static1.squarespace.com/static/5978a800bf629a80c569eef0/t/5aa99a6c9140b77920bc37d1/1530223323224/Toxic+Chemical+Reduction+in+Buildings.pdf [Accessed 15 Sep. 2023].
(78) Sustainable Concrete. (n.d). Available at: https://www.sustainableconcrete.org.uk/ [Accessed 15 Sep. 2023].
(79) Bai, Z. et al (2006). Emission of ammonia from indoor concrete wall and assessment of human exposure. *Environment International*, 32(3), 303–311. Available at: https://doi.org/10.1016/j.envint.2005.06.002 [Accessed 15 Sep. 2023].
(80) Johansson, I. (2014). 'Emissions from concrete-an indoor air quality issue?' [pdf] *AK-konsult Indoor Air AB*, pp. 330–37. Available at: http://media1.kumlin.biz/2016/02/Emissions-from-concrete.pdf [Accessed 15 Sep. 2023]
(81) Cormier, S.A. et al. (2006). 'Origin and health impacts of emissions of toxic by-products and fine particles from combustion and thermal treatment of hazardous wastes and materials.' *Environmental Health Perspectives*, 114(6), 810–817. Available at: https://doi.org/10.1289/ehp.8629. [Accessed 15 Sep. 2023].
(82) Klimenko, V. G., Kashin, G. A., and Prikaznova, T. A. (2018). 'Plaster-based magnetite composite materials in construction Prikaznova IOP Conf.' *Series: Materials Science and Engineering*, 327 (2018) 032029.
(83) Däumling, C. et al. (n.d). 'Emissions from Oriented Strand Boards (OSB) Covered with Gypsum Plates.'[pdf] *Sentinel Haus.* Available at: https://blog.sentinel-haus.eu/wp-content/uploads/2010/06/279-HB09-Publ-OSB.pdf [Accessed 15 Sep. 2023].
(84) Sterling OSB Material Safety Data Sheet. (2015). Revision 8 Available at: https://dam-assets.apps.travisperkins.group/R0vNP/GPID_1000071569_COSHH_00.pdf [Accessed 15 Sep. 2023].
(85) West Fraser (n.d). 'Norbord's Sterling OSB Zero Range-Zero-Added Formaldehyde (ZAF) For Safer Construction and Healthier Buildings.' Available at: https://uk.westfraser.com/news/norbords-sterlingosb-zero-range-zero-added-formaldehyde-zaf-for-safer-construction-and-healthier-buildings/ [Accessed 15 Sep. 2023]
(86) U.S Environmental Protection Agency. (2011). 'Methylene Diphenyl Diisocyanate (MDI) And Related Compounds Action Plan.' [pdf]. Available at: www.epa.gov/sites/default/files/2015-09/documents/mdi.pdf [Accessed 15 Sep. 2023]
(87) Health and Safety Executive (n.d). Medium Density Fibreboard (MDF). Available at: www.hse.gov.uk/woodworking/faq-mdf.htm#:~:text=The%20release%20of%20formaldehyde%20from,produce%20only%20low%20emission%20boards [Accessed 16 Sep. 2023]
(88) Direct Doors (n.d). Solid Wood Doors. Available at: www.directdoors.com/blogs/help/choosing-your-door-solid-wood-doors [Accessed 16 Sep. 2023].
(89) Corinne. (2019). 'Non-toxic interior and exterior doors: A guide to door types and materials and the chemicals they offgas.' *My Chemical-Free House.* Available at: www.mychemicalfreehouse.net/2019/08/non-toxic-doors-guide-to-door-types-and.html [Accessed 16 Sep. 2023].
(90) Fuglestad, F. E. et al. (2022). 'Determining the vapour resistance of breather membrane adhesive joints.' *Materials*, 15, 6619. Available at: https://doi.org/10.3390/ ma15196619 [Accessed 16 Sep. 2023].
(91) Ghomi, E. R et al. (2020). 'The flame retardancy of polyethylene composites: From fundamental concepts to nanocomposites' *Molecules* 25(21), 5157. Available at: https://doi.org/10.3390/molecules25215157 [Accessed 16 Sep. 2023].
(92) RIBA. (n.d). Are Your Wall Membranes Non-combustible and Smoke-blocking? Available at: www.architecture.com/knowledge-and-resources/knowledge-landing-page/are-your-wall-membranes-non-combustible-and-smoke-blocking [Accessed 16 Sep. 2023].
(93) Swedish Chemicals Agency (2015). SUBSTANCE EVALUATION CONCLUSION as required by REACH Article 48 and EVALUATION REPORT for Trimethoxy(vinyl)silane. Available at: https://echa.europa.eu/documents/10162/6709659e-4c4c-06a7-ea70-5aa094c25203 [Accessed 16 Sep. 2023].

(94) Serge Ferrari. (2020). Non-Combustible Facade Breather Membrane [pdf]. Available at: https://itpltd.com/wp-content/uploads/Whitepaper_Stamisol_Safe_One_UK_05_2020-1.pdf [Accessed 16 Sep. 2023].

(95) Tricarico, E. (2015). Where Your Money Goes: A Worldwide Tour of Rio Tinto's Wreckage. Available at: https://foe.scot/where-your-money-goes-a-worldwide-tour-of-rio-tintos-wreckage/ [Accessed 16 Sep. 2023].

(96) (2023) 'Substance Infocard', European Chemicals Agency. Available at: https://echa.europa.eu/substance-information/-/substanceinfo/100.014.129 [Accessed 16 Sep. 2023].

(97) Directorate-General for Health & Consumers. (2010). Opinion on Boron Compounds. Available at: https://ec.europa.eu/health/scientific_committees/consumer_safety/docs/sccs_o_027.pdf [Accessed 16 Sep. 2023].

(98) Pongsavee, M. (2009). 'Effect of borax on immune cell proliferation and sister chromatid exchange in human chromosomes.' *Journal of Occupational Medicine and Toxicology*, 4, 27. Available at: https://doi.org/10.1186/1745-6673-4-27 [Accessed 16 Sep. 2023].

(99) BORAX (2021). Borates for Fire Retardancy in Cellulosic Materials. [pdf] Available at: www.borax.com/BoraxCorp/media/Borax-Main/Resources/Technical-Bulletin/borates-fire-retardancy-cellulosic-materials.pdf?ext=.pdf [Accessed 16 Sep. 2023].

(100) Page, J. et al (2023). 'A new consensus on reconciling fire safety with environmental & health impacts of chemical flame retardants,' *Environment International*, 173, Available at: https://doi.org/10.1016/j.envint.2023.107782. [Accessed 16 Sep. 2023].

(101) House of Commons Environmental Audit Committee. (2019). *Toxic Chemicals in Everyday Life*. [pdf] Available at: https://publications.parliament.uk/pa/cm 201719/cmselect/cmenvaud/1805/1805.pdf. [Accessed 16 Sep. 2023].

(102) Howes, L. (2013). 'Uproar over chemical exposure advice for pregnant women.' *Royal Society of Chemistry*. Available at: www.chemistryworld.com/news/uproar-over-chemical-exposure-advice-for-pregnant-women/6243.article [Accessed 16 Sep. 2023]

(103) Bellingham, M. and Sharpe, R. (2013) 'Chemical Exposures During Pregnancy: Dealing with Potential, but Unproven, Risks to Child Health' *Scientific Impact Paper No. 37*. London: Royal College of Obstetricians and Gynaecologists.

(104) Official Journal of the European Union. (2003) '*Directive 2003/11/EC of the European Parliament and of the Council*' [pdf]. Available at: https://eur-lex.europa.eu/LexUriServ/LexUriServ.do?uri=OJ:L:2003:042:0045:0046:EN:PDF [Accessed 16 Sep. 2023].

(105) Leatham, X. (2021). 'Toxic flame retardants linked to infertility and hormone problems are discovered in the blubber of KILLER WHALES for the first time.' *Daily Mail*. Available at: www.dailymail.co.uk/sciencetech/article-9594975/Toxic-flame-retardants-discovered-blubber-KILLER-WHALES-time.html [Accessed 16 Sep. 2023].

(106) Agency for Toxic Substances and Disease Registry. (2017) *Polybrominated Diphenyl Ethers – ToxFAQs*. [pdf]. Available at: www.atsdr.cdc.gov/toxfaqs/tfacts207.pdf [Accessed 16 Sep. 2023].

(107) Huang, J. et al. (2022). 'Industrial production of Organophosphate Flame Retardants (OPFRs): Big knowledge gaps need to be filled?' *Bulletin of Environmental Contamination and Toxicology*, 108, 809–818. Available at: https://doi.org/10.1007/s00128-021-03454- [Accessed 16 Sep. 2023].

(108) Wang, X. et al (2021). 'Human internal exposure to organophosphate esters: A short review of urinary monitoring on the basis of biological metabolism research.' *Journal of Hazardous Materials*, 418. Available at: https://doi.org/10.1016/j.jhazmat.2021.126279 [Accessed 16 Sep. 2023].

(109) Safer States. (n.d). Toxic Flame Retardants. Available at: www.saferstates.org/toxic-chemicals/toxic-flame-retardants/#:~:text=Flame%20Retardants%20have%20been%20shown,levels%20of%20these%20toxic%20chemicals. [Accessed 16 Sep. 2023].

(110) Stec, A. A. and Hull, T. R. (2011). Assessment of the fire toxicity of building insulation materials. *Energy and Buildings*, 43(2–3), 498–506. Available at: https://doi.org/10.1016/j.enbuild.2010.10.015. [Accessed 16 Sep. 2023].

(111) Yates, T. (2017). 'Study to evaluate the need to regulate within the Framework of Regulation (EU) 305/2011 on the toxicity of smoke produced by construction products in fires.' *EU Commission*. Available at: https://ec.europa.eu/docsroom/documents/27346 [Accessed 16 Sep. 2023].

(112) Modern Building Alliance. (n.d). Smoke Toxicity & Smoke Management. Available at: www.modernbuildingalliance.eu/smoke-toxicity/ [Accessed 16 Sep. 2023].
(113) Stec, A. A. et al. (2018). 'Occupational Exposure to Polycyclic Aromatic Hydrocarbons and Elevated Cancer Incidence in Firefighters.' *Scientific Reports*, 8(1). Available at: https://doi.org/10.1038/s41598-018-20616-6 [Accessed 16 Sep. 2023].
(114) Firehouse. (2023). Many Grenfell Tower Firefighters Diagnosed with Incurable Cancers. Available at: www.firehouse.com/safety-health/cancer-prevention/video/21292566/many-grenfell-tower-firefighters-diagnosed-with-incurable-cancers [Accessed 17 Sep. 2023].
(115) Both, R. (2023). 'Firefighters three times more likely to die from certain types of cancer- study' *The Guardian*. Available at: https://www.theguardian.com/uk-news/2023/jan/10/firefighters-cancer-toxic-chemicals-study [Accessed 17 Sep. 2023].
(116) Agreed EU-LCI Values. (2021). Available at: https://ec.europa.eu/docsroom/documents/49239 [Accessed 17 Sep. 2023].
(117) Dodson, R. E. et al. (2017). 'Chemical exposures in recently renovated low-income housing: Influence of building materials and occupant activities.' *Environment International*, 109, 114–127. Available at: https://doi.org/10.1016/j.envint.2017.07.007 [Accessed 17 Sep. 2023].
(118) Huang, L. et al (2022). 'Chemicals of concern in building materials: A high-throughput screening.' *Journal of Hazardous Materials*, 424(Pt C). Available at: https://doi.org/10.1016/j.jhazmat.2021.127574 [Accessed 17 Sep. 2023].
(119) Weschler, C. J. (2009). 'Changes in indoor pollutants since the 1950s' *Atmospheric Environment*, 43(1), 156–169. Available at: https://doi.org/10.1016/j.atmosenv.2008.09.044 [Accessed 17 Sep. 2023].
(120) DRAX. (2017). Power Generation. Available at: www.drax.com/power-generation/wont-recognise-powder-will-know-used/ [Accessed 17 Sep. 2023].
(121) Allen, G. (2009). 'Toxic chinese drywall creates a housing disaster' *NPR*. Available at: www.npr.org/2009/10/27/114182073/toxic-chinese-drywall-creates-a-housing-disaster [Accessed 17 Sep. 2023].

3 Emissions from building materials and health impacts

At the heart of this book is the question of emissions from building materials. Are they a problem, do they even exist and why are they so readily dismissed by some scientists, academics and the construction industry, particularly in the UK? Buildings today are constructed of a wide range of materials, but most of them contain a significant amount of chemicals as has been illustrated in Chapter 2. Some of these may have been used in the past, but the addition of new chemicals has grown massively in recent years. These chemicals are used for a variety of reasons, to reduce costs, to speed up manufacture, to aid with drying or curing, to add to longevity (though this may not be a valid reason) and resistance to moisture and mould and for many other reasons. Some regard synthetic chemical-based materials as better than natural traditional ones while ignoring the issue of hazardous emissions.

It is perfectly possible to construct buildings without too many chemicals, but unfortunately the use of chemicals has increased significantly, as Weschler (2009) states. As he explains, these chemicals were not present 50 years ago but the numbers have increased even more significantly since he wrote this 15 years ago:

> Since the 1950s, levels of certain indoor pollutants (e.g., formaldehyde, aromatic and chlorinated solvents, chlorinated pesticides, PCBs) have increased and then decreased. Levels of other indoor pollutants have increased and remain high (e.g., phthalate esters, brominated flame-retardants, non-ionic surfactants and their degradation products). Many of the chemicals presently found in indoor environments, as well as in the blood and urine of occupants, were not present 50 years ago. The consequent changes in emission profiles for indoor pollutants have been accompanied by modifications in building operations. Residences and non-residences are less ventilated than they were decades ago. **(1)**

International research on emissions from chemicals in buildings is extensive but much less so in the UK where, for unexplained reasons, many academics and scientists, concerned with air quality (with a few notable exceptions), have decided to ignore this problem. They argue that the data on chemical emissions does not exist, though that is easy to say when they have not done any research into the problem. The science of chemical emissions is complex because manufacturers do not always disclose the chemical constituents of products. Even where products are registered with the Registration, Evaluation, Authorisation and Restriction of Chemicals (REACH register) and health and safety data sheets are available, it can take a Sherlock Holmes to identify and analyse the likely hazards. **(2)**

Chemicals can be emitted from the materials, when installed in buildings, at different rates. Some chemicals may be encapsulated by other materials and their contact with the indoor air can vary. The rate of off-gassing of the chemicals can vary with some emitted at high concentrations,

DOI: 10.1201/9781003129226-3

when first applied, but then diminish over time. Others can linger in a building for many years and some can react with other chemicals such as ozone and change into different chemicals. Data on emissions may be available from laboratory studies that have been carried out in a range of climate chambers, some the size of a domestic microwave, others the size of a room. How the emissions in actual buildings may be very different from the laboratory studies. Measurement of indoor air emissions using air and dust sampling, as discussed in Chapter 7, are useful tools and can identify a range of hazards.

The UK Government has a poor record on the health effects of chemicals as it has not published a new Chemicals Strategy since 1999. A 25-year Environment Plan published in 2018 committed the UK to a new UK Chemicals Strategy but work on this was paused during the Covid pandemic. The UK has introduced a Chemicals Management regime to replace REACH when the UK left Europe but campaigning organisation Chem Trust suggests that people and the environmental in the UK will be less protected from chemicals than in Europe. **(3)**

Off-gassing or out-gassing

The issue of "off-gassing" of chemicals from materials is an area of much debate. While some scientific work has gone into measuring off-gassing rates, some academics and scientists have assumed that it is not a long-term problem. It is assumed that emissions of chemicals will be much higher in the short period, after construction or decorating work, and that these will gradually decrease over time to acceptable levels. This is an assumption that underpins many policies such as building regulations and health policies.

Insufficient scientific work has been carried out on off-gassing, and the literature focussed on emissions from materials has largely been carried out using climate chambers rather than real buildings. This work in climate chambers can provide useful but sometimes limited data and while there is some data available it is possible to draw conclusions as to which materials and chemicals present a higher risk.

The rate of decay of these emissions is not sufficiently well documented, but it is not sensible to dismiss emissions as a short-term problem following recent application of materials. Many of the chemicals are persistent and remain in buildings long after they were installed, though much more research is needed to determine rates of off-gassing, as the data currently available is limited to a small proportion of chemicals.

> While VOC emissions from a product tend to decrease sharply during the first weeks or months of a product's life, SVOC emissions tend to continue throughout the life of a product. Additionally, SVOCs sorb to other indoor surfaces after they are released to the air, and, in the case of flow volatility SVOCs (e.g., DEHP, BDE209, DDT), these sorbed SVOCs continue to desorb long after the host material is removed.
>
> Given the importance of this subject, it is striking how little directly measured, year-to-year data exist on the kind and concentration of indoor pollutants). **(4)**

A further problem is presented when attempting to assess the health risks to building occupants from chemical emissions. Epidemiological data is sparse and is often based on assumptions about hospital and medical admissions, where little if any attempt has been made to correlate this with the buildings that patients have been occupying. ASHRAE in the USA has attempted to assess emissions problem and set guidelines for safe levels of off-gassing, to guide engineers designing ventilation systems. This is based on the assumption that mechanical ventilation is the main way to remove hazardous emissions and maintain an acceptable level of indoor air quality. There is no doubt that continuous high-level ventilation will dilute the concentration of indoor pollutants, but it does not remove them entirely. Some of the ASHRAE literature also refers to *source control* which is recognised as another method of limiting emissions. Source control simply means assessing materials to be used in a building and reducing or excluding those that may be responsible for higher emissions.

Building emissions deniers ("BEDS")

When beginning research into this topic in 2015 it was immediately clear from the international body of literature on indoor air quality that emissions from building materials was a major concern, but the majority of the literature on this was from outside the UK. Scientists in the USA and some European countries have paid much more attention to the issue of emissions from building materials. When raising this issue with scientists, academics and policy makers in the UK, it is common to be rebuffed with statements that this is not a major concern. Evidence of this is discussed in detail in Chapter 4. Building materials, we are told, are already well regulated. When arguing against this, building emissions deniers (BEDs) often claim that there is insufficient evidence of emissions from building materials. There is no doubt that evidence in the UK is in short supply as scientists working on air quality had not carried out much research to identify such emissions, or they have chosen in advance what emissions to detect, such as from cooking or wood burning stoves. It is easy to use the argument that there is a lack of evidence when little work has been carried out to find the evidence! When confronted with available evidence of emissions from materials the next denial argument is that there is no evidence that these emissions have any negative health impacts. Once again, the problem is that very little research has been done in the UK either in the fields of medical epidemiology or emissions science. A leading air quality expert stated at a meeting that building materials were already well regulated (probably thinking of the reduction of solvents in paints), but in reality, VOC and other emissions in buildings is poorly regulated. It is easy to deny that a problem exists if research into it has been ignored. Considerable effort has gone into claims that the main indoor air quality problems are a result of traffic pollution, cooking, emissions from wood burning stoves and hygiene products. For instance, the UK Clean Air programme has been almost entirely focussed on ambient air and traffic pollution, though some recent research under this programme touches on indoor air quality as discussed in Chapter 4.

LCI database

A credible source of information for emissions from chemicals used in building materials is the LCI database established by the EU. **(5)**, **(6)**

> It was recommended that "lowest concentrations of interest" (LCI) should be derived on the basis of either air quality guideline values (AQG) or occupational exposure limits as auxiliary parameters for the assessment of the health risks resulting from exposure to chemicals emitted from building materials. The concept of LCI values was adopted and further developed/revised in the German AgBB scheme and in the French AFSSET (now ANSES) protocol for the health-related evaluation of VOC emissions from building products.

When accessing the LCI list on the internet it is possible to find detail fact sheets on each substance listed which can be downloaded.

> EU-LCI values have been developed for the evaluation of single construction product emissions and they do not constitute guideline values for indoor air quality. The data compiled during the setting of EU-LCI values may be informative for the process of establishing guidelines by other expert groups but has been evaluated by the EU-LCI Sub-Group according to the published protocol (ECA report No, 29) concerning emissions from construction products.
>
> EU-LCI are thus based on reported scientific data and expert judgment and represent concentration levels that are considered likely not to cause adverse effects over the longer term by use of the model room as a reference.
>
> EU-LCI values are used for assessing single product emissions after 28 days during a laboratory test chamber procedure. They are applied within health-related evaluation schemes to assess health risks from indoor product emissions on the basis of life-long exposure. **(7)**

EU-LCI values are health-based reference concentrations of chemical substances for inhalation exposure used to assess emissions after 28 days from a single construction product during a laboratory test chamber procedure as defined in the EN 16516 standard. EU-LCI values are applied in product safety assessment with the ultimate goal to avoid health risks from long-term exposure of the general population. They are expressed as µg/mÑ.

EU-LCI values have been developed for the evaluation of single construction product emissions and do not constitute guideline values for indoor air quality. The data compiled during the setting of EU-LCI values may be informative for the process of establishing guidelines by other expert groups but has been evaluated by the EU-LCI Sub-Group according to the published protocol (ECA report No, 29) concerning emissions from construction products.

EU-LCI values are derived using a compilation of epidemiological and toxicological data from risk assessments published by established international and national committees and/or other relevant studies. EU-LCI are thus based on reported scientific data and expert judgment and represent concentration levels that are considered likely not to cause adverse effects over the longer term by use of the model room as a reference. The derivation of EU-LCI values involves three main steps: compilation of toxicological data, data evaluation and derivation of the EU-LCI value on the basis of a total (combined) assessment factor, based on established risk assessment principles and expert judgment, laid out in a factsheet.

EU-LCI values are used for assessing single product emissions after 28 days during a laboratory test chamber procedure. They are applied within health-related evaluation schemes to assess health risks from indoor product emissions on the basis of life-long exposure. EU-LCI values are related to similar “emission limit” values published by other authorities such as Anses (former AFSSET) in France or AgBB in Germany.

EU-LCI, French CLI and German NIK values have the same definition but historically different derivation procedures. In order to support the harmonisation of the health-based evaluation of construction product emissions in Europe. The AgBB has, since 2015, adopted the EU-LCI values and their derivation procedure.

The LCI database is a tremendous resource in terms of understanding the risks of emissions from chemicals in buildings, but it has its limitations in that it cannot entirely predict emissions levels in actual buildings. This can only be determined by air sampling or other evaluation measures on site. However, what the LCI database does do very effectively is give the lie to the assertion by BEDs that there is no evidence of emissions. It is noteworthy that no reference to the LCI database was made in either the NICE or DEFRA AQEG reports about indoor air quality discussed in Chapter 4, thus ignoring an important piece of research work on building material emissions even though a leading figure in the development of LCI was Derrick Crump, chair of the UK Indoor Environment Group (UKIEG) who previously worked at the UK Building Research Establishment and Cranfield University.

To illustrate the value of the LCI database two examples downloaded from the data base are discussed in more detail.

Toluene

Toluene is a petrochemical derived product used mainly as a solvent in paints, paint thinners, silicone sealants, adhesives, rubber, lacquers and disinfectants. Inhalation of toluene in low-to-moderate levels can cause tiredness, confusion, weakness, drunken-type actions, memory loss, nausea, loss of appetite and hearing loss. Toluene is used in "glue sniffing" for its intoxicating effect. A 13-page document that can be downloaded from the LCI database cites 10 academic references to the health effects of emissions from Toluene emphasising its neurological effects. It is possible to identify paints and other building products that include Toluene from health and safety data sheets.

Styrene

Styrene is flammable, is a skin and eye irritant and is acutely toxic if inhaled. It has neuro-psychological effects. The LCI lists 14 references and includes the following:

> According to the Swedish consensus report (Montelius 2010), the critical effects of styrene are genotoxicity, hearing loss and effects on color vision. Styrene is probably genotoxic to humans and possibly also carcinogenic. Genotoxic effects (chromosomal aberrations, micronuclei, strand breaks, DNA repair) have been observed at occupational exposures down to about 10 ppm.

The LCI database factsheet Styrene can be accessed through the LCI database. **(8)**

Does ventilation reduce off-gassing?

Ventilation is discussed in more detail in Chapter 6. Chemicals in buildings possibly can be reduced to an acceptable level through ventilation particularly in the first two years. Assumptions have been made that after this initial period, emissions have declined to a level that is not a risk to health. It is even suggested by a number of experts that buildings should be left unoccupied for 6 months and that emissions should be "baked off" by overheating the building to 30 degrees C because some VOCs will be given off at higher rates at higher temperatures. **(9)**

Those exponents of the ventilation solution to indoor emissions rarely, if ever, question whether we should be using all these chemicals, as it is assumed that they are a normal part of building construction. Ventilation as a way to reduce emissions imposes a higher energy burden on buildings, baking off, much increased ventilation and leaving buildings empty for half a year. Nor surprisingly this is not often done!

However, there are some fundamental questions hidden away here. Firstly, the assumption that emissions will decay to acceptable levels over time. The building emissions deniers (BEDs) rely heavily on this assumption and refer to what is called the "off-gassing phase" as though this is time limited. Most of the scientific studies on off-gassing rely on Total Volatile Organic Chemical (TVOC) analysis, such as Holøs et al, and rarely distinguish between the many hundreds of chemicals and toxic substances that might be present, all of which will off gas at different rates. They build in an assumption that 200ug/m3 as an acceptable TVOC level. In fact, 200 is lower than the level set in the UK building regulation or other standards. Also, there has been very little research into the reaction of different people to off-gassing chemicals. Some people may be hypersensitive to much lower levels of chemicals than others. The chemicals lingering in buildings for many years may continue to have negative health effects on some people whereas others may not be affected. **(10)**

There is no doubt that ventilation, particularly forced ventilation, can reduce concentrations of chemicals in the air, but it doesn't necessarily remove the source. Some chemicals used in buildings are scientifically described as persistent; the term forever chemicals is now well known, and the absorption of chemicals into the body through different means, thyroid, lungs may continue over many years, irrespective of ventilation levels.

How to measure off-gassing and its health effects

Off-gassing, or "out-gassing" as it is sometimes referred to, still requires considerable scientific research in order to prevent it from being dismissed as a source of hazardous building emissions. Huang et al proposed an important methodology to measure this in 2019, but it is very complex and there is little evidence of their proposals being adopted so far. As they state,

> there do not currently exist scientifically defensible ways to consistently characterise the human exposures to near field chemical emissions and associated health impacts during the use stage of building materials. **(11)**

The approach which they propose is based on the work of Jolliet who defined the concept of "product intake fraction" (PiF). Jolliet et al have considered the impact of more than 80,000 chemicals in "consumer" products. They claim that 9% of the chemical-product combinations have lifetime cancer risks though they don't distinguish construction products in their mass balance study. **(12)**, **(13)**

Jolliet and his colleagues have produced a methodology for assessing materials emissions that they say can be included in Life Cycle Analyses which are commonly required for Environmental Product Declarations, published for many building materials. Standard LCA methodology includes assessing the impact of chemicals in products on the wider environment, but as Huang and Jolliet point out, while information on chemical content in LCAs and EPDS exists, the impact of these on health is not measured. They are highly critical of a number of key books and reports on the life cycle and environmental impact of materials that do not consider human exposures and health impacts, including the highly influential 2009 book, *Berge Ecology of Building Materials*. **(14)**

Despite the brilliance of Huang and Jolliet's work they may be overly optimistic about the likelihood of their proposals being adopted by mainstream construction materials companies. They do list a number of American materials guides for green buildings, some better than others, but some hover on the edge of greenwash. Examples of these such as Health Product Declarations (HPDs) are more focussed on marketing products than providing serious warnings about health risks. A proposal to introduce HPDs into Europe was blocked by a working group dominated by petrochemical manufacturing businesses who do not appear want to pursue the idea. The HPD collaborative in the US lists Saint Gobain and DuPont as major collaborators. **(15)**

Huang and Jolliet emphasise the impact of increased energy efficiency and air tightness in buildings in recent decades, and the increased use of synthetic materials and the resulting impact on increased exposures to hazardous chemicals and the absence of emissions standards for chemicals in building materials except for a few exceptions such as formaldehyde. In order to develop their methodology Huang and Jolliet first explain the variations in emissions rates. In order to explore this, they go back to literature from 1994, 2001 and 2004. Surprisingly little research has been carried out on release mechanisms, emissions rates and decay rates for VOCs and other chemicals in more recent times. What little research has been done indicates a variation in emissions and decay rates. Not all chemicals are released or decay at similar rates. As has been explained, while some dissipate very early in their life cycle, others can take much longer.

Having established that chemical emissions from building materials can be a problem, the next question is that of human exposures to chemicals. Chemicals can be deposited on surfaces and absorbed by furnishings. Concentrations of chemicals can vary from room to room with the kitchen having the highest risk according to Carter et al. **(16)**

Exposure of the human body to chemicals

There are various methods of determining exposure of the body to chemicals and how they are absorbed and this varies with different chemicals and whether they are absorbed through the skin or breathing. Huang and Jolliet indicate that this is an area that needs further research. We return to this in Chapter 5. There is a range of literature on toxicological effects and human health impacts with various methods that have been developed including *Usetox* which has been developed by SETAC to characterise the human toxicological impacts of chemicals emissions in lifecycle assessment (SETAC is the Society of Environmental Toxicology and Chemistry). **(17)**

The abstract book for the SETAC 33rd annual meeting held in Dublin in 2023 is 873 pages long but only two papers seemed to address the issue of construction materials.

Huang and Jolliet point out that tens of thousands of chemicals are used in building materials, but only a small number have been identified in scientific and healthy building literature. For instance, they claim that the American Pharos database identifies 756 chemicals of which only 184 are listed by *Usetox*. However, Pharos do provide a database called Green Screen of over 182,265 chemical hazards including a list of 229 different kinds of building products. The Pharos project is part of the Healthy Building Network which provides a useful source of information on building products. They are listed in colour with the greenest in green down to the most negative in dark red. The inclusion of Fibreglass products in the green section is surprising and indicates the dominance of Glass Fibre in the USA. **(18)**, **(19)**

Huang and Jolliet propose a framework to assess the chemical exposure and health impacts of building materials based on the work of Fantke. Their aim is to use Life Cycle Analysis (LCA) methods to combine the amount of materials used in terms of what is called a Functional Unit (FU) in a Life Cycle Inventory (LCI). **(20)**, **(21)**

Huang and Jolliet introduce another term PiF. The Product Intake Fraction:

> ...defined as the fraction of a chemical used in a product application that is cumulatively taken in by the user and by the general population during use and disposal stages. Note that PiF includes exposures during both the use and disposal stages. **(22)**

This methodology is concerned with the exposure during the use phase of materials in a building. Chemicals can be absorbed from near-field sources as well as far-field such as ambient air.

Table 3.1 Pharos Healthy Building Network list of Insulation materials

Dark Green	Expanded Cork Board
	Blown-in Wood Fibre Fill
Light Green	Blown and Batt Sheep's Wool
	Hemp Fibre Batts
	Wood Fibre Batts and Boards
	Blown-in Fibreglass or Mineral Wool
	Unfaced Fibreglass Matts
	Blown in Cellulose (Loose Fill)
	Unfaced Formaldehyde-Free Fibreglass Boards
Yellow	Kraft-Faced Fibreglass batts
	Cellulose/Cotton Batts
	Blown in Cellulose (Dense Pack and Wet Blown)
Orange	PSK or FSK Fibreglass Batts
	Standard Mineral Wool Batts and Boards
	Standard Fibreglass Boards
	Halogen Free Polyisocyanurate Boards
Red	ASI or FSK Fibreglass Boards
	Expanded Polystyrene (EPS) Boards
	Standard Polyisocyanurate Boards
	Extruded Polystyrene Boards
Dark Red	Spray Polyurethane Foam

Human intake can include respiratory tract, gastrointestinal tract, and epidermis (skin). Fantke also introduces the concept of TFs (Transfer Fractions) which define the cumulative intake that is then related to the Functional Unit (FU) used to measure the weight of materials in a standard LCA.

> The resulting cumulative transfer fractions from the compartment of entry to various human intake compartments are thus PiFs by different exposure routes. (Fantke et al. 2016) Multiplying the PiFs by the amount of chemical used per FU determined in the LCI phase yields the intake per FU as output of the exposure assessment. This framework (Fantke et al. 2016) allows the determination of both a per functional unit approach and a more risk assessment-oriented approach based on individual doses expressed in milligrams per kilogram of body weight per day or in kilograms per person per day. **(23)**

This is made to look more complicated due to the determination of Hang and Fantke to relate measuring human intake of chemicals to standard LCA methods. They then suggest that using measured intake doses they can assess the disease severity expressed in DALYS which are Disability Adjusted Life Years.

> DALYs are calculated as the sum of years of life lost to premature mortality and morbidity in a population for some negative health effect.
>
> A second method might be to identify properties of a building that are known to affect IAQ directly, for example using a tick-box approach. Each feature could be weighted according to their hazard and aggregated to produce a single metric. This method could be used to develop a third-party rating system, similar to many existing energy rating schemes, and might be helpful to those sensitive to specific contaminants.

> We use 827 data sets to evaluate the concentrations of 45 airborne contaminants in dwellings The five most abundant, by mass, are ethanol, PM10, formaldehyde, PM2.5 and nitrogen dioxide (NO2) To include the harm budget approach in relevant standards, several key aspects should be considered: (i) limiting the contaminants of concern to two or three of the most harmful would be beneficial to make source control, remediation, diagnostics, and enforcement simpler, and (ii) instead of using absolute terms, it would be useful to consider the harm budget in relative terms, using a dimensionless magnitude. **(24)**

There appear to be dangers in the HARM/DAYLS method of analysing the hazards from indoor air contaminants through over-simplifying the problem, and these issues were raised with Dr Benjamin Jones, one of the authors of the Morantes paper.

> Any contaminants that do not appear on the list are either less harmful than those we present or there is insufficient data to consider them.
>
> Some people are always more sensitive than others. However, we are considering populations of people and not individuals.
>
> We do consider the harm from mold, but find it less harmful than 6 other contaminants, and it contributes significantly less than 1% of population harm from indoor contaminants. The spelling of mold reflects the American audience we aim to appeal to because the work will be included a new addendum to ASHRAE Standard 62.2 that would add a harm-based Indoor Air Quality procedure as an alternative compliance method. In addition to the peer review, the work is also being assessed by ASHRAE and IEA Annex 86. It is currently out for public review in the US. **(25)**

The assertion that mould contributes less than 1% of harm from indoor contaminants will come as a shock to the hundreds of thousands of people living in fuel poverty in the UK and battling with mould contamination, which has even forced the conservative Government to strengthen legislation. This provides a warning that there may be flaws in such a methodology.

But nevertheless, the Huang and Jolliet team may be moving in the right direction in their attempts to merge a widely accepted medical assessment tool of harm with LCA methods.

> It should be noted that our proposed systematic approach to assess the use stage impacts is different from the traditional LCA methods which have focused on manufacturing and distribution phases. While traditional LCA methods calculate the inventory as the amount of chemicals emitted to the environment during the considered life cycle stages, our proposed approach uses the amount of chemicals initially present in the product as the inventory output for the use stage. This inventory can be easily harmonized with the traditional LCI as they are both linked to the functional unit of the product. The impact assessment for the use stage thus includes both the chemical releases from the product and the resulting human exposures, which are represented by the PiF values. **(26)**

Assessing human health damage from indoor exposure to chemicals emitted from building materials is not an easy task. An important recent paper examines the impact of three common insulation materials: Expanded polystyrene (EPS), Extruded polystyrene and Polyurethane (PU). **(27)**

One of the reasons that emissions from building materials are ignored or swept under the carpet is due to the lack of published data on emission levels. Maury-Micollier et al have compared the available data on chemical contents and related it to health. Hazardous emissions from insulation materials are frequently ignored as manufacturers will argue that insulation materials are encapsulated within building structures (often for fire safety reasons) and thus emissions from these materials are claimed to be very low or almost non-existent.

An example of a denial of hazardous emissions was found on the PU Europe website.

This body representing the manufacturers of plastic foam insulation products (a £60 billion industry in Europe) claim that

> "Once the foam has cured, as with other forms of PU used in buildings[...], it is considered chemically inert. VOC and SVOC emission levels are then comparable to those of factory-made PU insulation." PU insulation is considered a very low emission product. In fact, emissions from PU products are well below those of most other insulation products. In particular, natural insulants can have VOC emission levels more than 100 times higher than those of PU. Very importantly, no carcinogenic, mutagenic or reprotoxic substances were detected in any of the emission tests on PU foam (See section 9 (Nachweise) in the EPDs at http://bau-umwelt.de/hp545/Daemmstoffe.htm) **(28)**

The absurd statement that natural insulants can have VOC emissions levels more than 100 times higher than those of PU is based on a reference that does not exist in which they say "See sections 9 (Nachweise) in the EPDs at http://bau-umwelt.de/ hp545/Daemmstoffe.htm."

The PU Secretariat replied to an enquiry about this and admitted that the Bau-Umwelt reference no longer existed (if it ever did), but this claim about emissions from natural materials remained on their website. PU insulation uses various blowing agents including Pentane which is a highly volatile compound. PU insulation (PUR/PIR) is produced by a reaction of diisocyanates (MDI) with polyols. MDI (methylene diphenyl diisocyanate) is a respiratory sensitizer suspected of causing cancer.

There is no doubt that the bulk of chemical emissions from plastic foam insulations occurs early on in the process. The chemical emissions can be detected by strong chemical smells if you enter a curing shed at one of the PU or PIR factories. Foam insulations need to cure to prevent shrinkage, but there is no doubt that shrinkage also occurs when the product is placed into buildings as discussed in Chapter 2. Once shrinkage occurs then gaps are left making the insulation almost worthless. As postoccupancy evaluation is rarely undertaken the drop in thermal performance is rarely measured. The question remains whether low-level emissions from the remaining insulation over a number of years (and its many chemical constituents) can have health effects. For many people, it may have no effect but for others, they can be hypersensitive to the chemicals and flame retardants that continue to emit (even at low levels though this is a controversial issue). **(29)**

There was a great deal of discussion of sick building and sick house syndrome 20 plus years ago such as a Japanese study published in 2000 which identified harmful substances emitted from building materials. **(30)**

Nordin in Sweden has recently published on the problem of multiple chemical sensitivity among some people. It is suggested by Nordin that the problem may be more psychological in nature rather than due to "toxic exposure," but he describes neurogenic inflammation as a genuine health problem, which can be caused by chemical irritants and asthmagens. He introduces the concept of nocebo (the opposite of placebo) in which the anticipation of exposure can cause a medical reaction. Thus, it is very important to have reliable data on chemical emissions from materials and in buildings and their health effects to distinguish real causal links. **(31)**

Table 3.2 Extract from Maury Micolier PU chemicals

TCPP (flame retardant)
Glycexrol
Formaldehyde (foaming agent)
MDI (isocyanate)
Bis ether (catalyst)
2 Butoxtethanol (plasticiser)
Dichlorethylne
Triethyl phosphate (plasticiser)
Pentamethyldiethylenetriamine
Ethanol (catalyst)
Dimethylionethanol (catalyst)
Trimethyl triazundecane
Propylene carbonate (structurant)
Ethylene glycol (antifreeze)

Maury Micolier, Huang and Jolliet, in their detailed analysis of emissions from three insulation materials, including PU, refer to numerous studies which indicate that in cured PU, the residual MDI weight fraction ranges from 1ppm to 14 ppm monomer, citing Hoffman et al. **(32)**

While these quantities might seem very low, PU insulation also contains flame retardants and formaldehyde. A table in the Maury Micolier paper lists 14 chemicals in PU foam insulation. **(33)**

These chemicals can be emitted through diffusion through walling materials. One pathway for external insulation is into external air which then gets into the house. Mawditt, who has extensively documented retrofit work to his own home, detected indoor air emissions from chemicals which must have come from the external wall insulation he had installed. **(34)**

It is not unusual to hear the argument that chemical emissions are prevented from affecting indoor air quality as materials such as plasterboard, plaster and concrete act as barriers, but in fact most building materials are vapour and gas permeable to some extent. Not only do Maury Micollier et al show that emissions can pass through common building materials, they point out that chemicals in concrete can also contribute to hazardous emissions. Common wall buildups consist of a range of materials such as concrete blocks as well as insulation. Most concrete now contains chemical additives such as Calcium Stearate to control efflorescence and Triethanolamine, as a water reducer, which are also hazardous chemicals that can be detected inside buildings.

A surprising finding of the paper is that formaldehyde, in particular, and other chemicals can diffuse through concrete:

> The diffusion of formaldehyde in concrete is so fast that even if the insulation is placed after a 20 cm-thick concrete layer, this substance will go through this layer **over 50 years** and reach the indoor air compartment. **(35)**

On the other hand, they say that concrete can significantly inhibit the transmission of flame retardants like tris(chloropropyl) phosphate (TCPP) if the insulation is on the outside of a concrete wall but increasingly foam insulation materials are placed on the inside of walls.

Maury Micolier et al have attempted to quantify off-gassing from construction consisting of five materials – paint, gypsum board, insulation, concrete and render – using a heat and mass transfer model. Chemical emissions can be transferred to and "sorbed" by different building elements. Relative humidity may also play a role on emissions rates, and there can also be chemical reactions within building materials. They point out that long-term experiments investigating these complex emissions issues are rare with the main focus having been on formaldehyde emission problems from composite wood products such as by He et al. **(36)**

Maury Micolier et al conclude that formaldehyde diffuses much faster than the TCPP flame retardant and state that it can migrate through concrete. The TCPP has a lower release rate. They state that the indoor air concentration of the formaldehyde (VOC) peaks after two weeks, and half of it is already reached over 5 years, whereas the TCPP (SVOC) flame retardants peaks only after two years and remains high over a 50-year period.

Given the serious concerns about flame retardants discussed, this is a worrying finding. They state that over 50 years, 0.4% of the formaldehyde initial mass has been taken in by occupants, whereas only 0.04% has been taken in from the TCPP. While this might suggest that the risk from flame retardants is lower, the fact that the chemicals linger in the building throughout its life is of concern. Maury Micolier calculate the mass of chemicals in products. The TCPP flame retardants in PU insulation is the highest containing a massive 21% of the total mass. Mass emissions from various products and chemicals varies significantly. Human health damage from PU insulation is due to formaldehyde and TCPP. MDI is also regarded as possible carcinogenic and can cause sensitisation and asthma.

The blowing agents in extruded polystyrene insulation (XPS) and expanded polystyrene insulation (EPS) have risks for health damage. Tetrafluoroethane using in XPS and Pentane in EPS can have health effects. **(37)**

The methodology using Product Intake fractions (PiFs) shows that these can vary considerably but how do they affect occupants? For most chemicals, inhalation is the dominant intake pathway. There can also be dermal gaseous uptake. There are numerous studies of dermal intake such as by Pelletier et al. 2017. **(38)**

Absorption can vary between different chemicals with different molecular weights. Dust ingestion and direct dermal contact pathways are predominant where concentrations at the materials surfaces of the building envelope or where chemicals emitted into the indoor air have been absorbed by the floor surface. Detecting chemical emissions has to involve air and dust sampling. The effects of chemicals on health vary from one material to another. Formaldehyde and Styrene have high carcinogenic effects and styrene in particular is easily metabolised in humans. The International Agency for Research on Cancer (IARC) classifies Styrene in Group 2A as probably carcinogenic to humans.

The polystyrene industry claims to use CO2 as a blowing agent, but pentane in particular has a very significant and distinctive smell and is easily detected in recent EPS products.

Maury Micollier warn that internal PU insulation has the most serious health impacts compared with EPS/XPS. The thickness of the insulation used must also be taken into account due to the greater mass of chemical constituents. They point out the health benefits claimed for better insulated buildings where thicker insulation is used must be seen as offset by the health damage due to the chemical emissions, particular with internal insulation. They conclude that "human health damage due to indoor exposure is dominant over outdoor exposure by two orders of magnitude for all insulation materials except XPS due to its use of Tetrafluoroethane".

Evidence of health effects from emissions from building materials

Two of the main chemicals that are readily emitted within buildings and have serious health concerns are PFAS and flame retardants.

PFAS

PFAS, Per- and polyfluorinated alkyl substances, represent a much larger health and pollution problem beyond concerns about indoor air. The societal cost of using toxic PFAS or "forever chemicals" across the global economy totals about $17.5tn annually, according to a study by Chemsec.**(32)**, **(33)** Chemsec, a Swedish NGO, has published several studies on the dangers of PFAS and lists the world's top 12 PFAS producers as AGC, Arkema, Chemours, Daikin, 3M, Solvay, Dongyue, Archroma, Merck, Bayer, BASF and Honeywell. PFAS are of particular concern because they are so prevalent in a wide variety of areas of human exposure, they are found in air and in dust inside buildings. **(39)**

Adverse health effects of PFAS are claimed to include, kidney and testicular, cancer, elevated cholesterol, liver disease, decreased fertility, thyroid, problems, and many other health problems. Biomonitoring efforts have revealed that a number of PFAS are present in the blood of European citizens, and although such levels of the most prevalent and well-studied PFAS like perfluorooctanoic acid (PFOA) and perfluorooctanesulfonic acid (PFOS) are declining, levels of newer generation PFAS are rising. Denmark, Germany, the Netherlands, Sweden and Norway announced that by July 2022 they were to formally propose to the European Chemicals Agency (ECHA) that these chemicals be restricted under Reach. Though it wasn't possible to discover the outcome of this.

> A comprehensive restriction is proposed, which includes a ban on the consumer use of substances, mixtures and articles as well as non-essential professional and industrial uses, as all use of PFAS potentially contributes to the accumulation of these extremely difficult-to-degrade chemical compounds in the environment. **(40)**

The main use of PFAS that affect indoor air in buildings are PTFE, Efte, plastic coating of wires, tapes in electrical and plumbing work, a whole range of composite wood products, artificial turf (now used indoors) carpets, hard flooring materials, windows, doors, glass, caulks and sealants, adhesives. PFAS are also used in a massive number of external materials, roofing materials, solar panels and so on. **(41)**

Much of the research on the health effects of PFAS has focussed on exposure through food and water, clothing and fabrics. Recognition of the high levels of PFAS in buildings materials is very new and it may be a while before the impact of this is fully understood. It will not be easy to avoid the use of materials containing PFAS as they are not always declared by manufacturers. **(42) (43)**

PFAS were recently in the news in the island of Jersey. Residents, who have suffered from cancer, claim that PFAS from the use of fire extinguishers at Jersey airport contaminated their water supply from bore holes, 30 years ago. The Island's Government is said to have signed confidential deal with 3M, the manufacturer of the foam towards a clean-up. Part of the deal was an agreement not to pursue any citizen legal actions. What is significant about this story, which does not affect indoor air quality, is the fact that PFAS have been around for 30 years. Similar problems have also involved legal action in Guernsey. **(37)**

International research has convincing evidence of the health dangers of PFAS and the US EPA has said that PFAs can cause serious health problems. For example, chemical company DuPont settled a £526 million lawsuit, the subject of a book by lawyer Roger Billot and a Hollywood film. Despite taking blood samples from islanders which showed elevated levels of PFA compounds Jersey's director of public health stated that this is a bit of a grey area! **(44) (45)**

PFAS in building materials and indoor air

> Sources of exposure to per- and polyfluorinated alkyl substances (PFAS) include food, water, and, given that humans spend typically 90% of their time indoors, air and dust. Quantifying PFAS that are prevalent indoors, such as neutral, volatile PFAS, and estimating their exposure risk to humans are thus important. To accurately measure these compounds indoors, polyethylene (PE) sheets were employed and validated as passive detection tools and analysed by gas chromatography–mass spectrometry. Air concentrations were compared to dust and carpet concentrations reported elsewhere. Partitioning between PE sheets of different thicknesses suggested that interactions of the PEs with the compounds are occurring by absorption. Volatile PFAS, specifically fluorotelomer alcohols (FTOHs), were ubiquitous in indoor environments. For example, in carpeted Californian kindergarten classrooms, 6:2 FTOH dominated with concentrations ranging from 9 to 600 ng m^{-3}, followed by 8:2 FTOH. Concentrations of volatile PFAS from air, carpet, and dust were closely related to each other, indicating that carpets and dust are major sources of FTOHs in air. Nonetheless, air posed the largest exposure risk of FTOHs and biotransformed perfluorinated alkyl acids (PFAA) in young children. This research highlights inhalation of indoor air as an important exposure pathway and the need for further reduction of precursors to PFAA. **(46)**

Paints are likely source of PFAS but only limited research has been done on the concentrations and emissions. Floor waxes and stain guards also have high use of PFAS.

> Different commercial paints ($n = 27$) were collected from local hardware stores and analysed for volatile PFAS by gas chromatography–mass spectrometry (GC–MS), non-volatile PFAS by liquid chromatography-quadrupole time-of-flight mass spectrometry (LC-qTOF), and total fluorine by ^{19}F nuclear magnetic resonance spectroscopy (NMR).
>
> Only 6:2 FTOH (0.9–83 μg/g) and 6:2 diPAP (0.073–58 μg/g) were found in five exterior and nine interior paints and only accounted for a maximum of 17% of total fluorine. one commercial paint exceeded the chosen reference dose (5 μg/kg-day) for children and adults, indicating the potential for human exposure during painting. **(47)**

The Guardian revealed that two major PFAS producers had hidden studies that suggested that the compounds are highly toxic at low doses in lab animals and stay in animals' bodies for much longer than was previously known though this was for packaging rather than building materials. **(48)**

PFAS are linked to kidney disease, cancer, neurological damage, developmental problems, mottled teeth and autoimmune disorders, while researchers also found higher mortality rates among young animals and human mothers exposed to the chemicals.

The Green Science Policy Institute identified the use of PFAS in a wide range of roofing materials and weather proofing membranes. PTFE and EFTE roofing materials and coatings are popular with architects, used in retractable and transparent roofs such as at the Eden project in Cornwall. **(49)**

Paints, plastic and metal coatings and a wide range of flooring materials all contain PFAS as well as a wide range of sealants, caulks and adhesives. PFAS are used in glass products, fabrics, wires and cables and tape in electrical and plumbing work. PFAS have become common in composite timber products and finally plastic grass. **(50)**

Flame retardants

The issue of flame retardants is discussed in Chapter 2. One of the ways in which flame retardants and other chemicals can affect the body is through endocrine disruption.

Endocrine disruption

There are many ways that chemicals in buildings can affect health, particularly those that pass through the thyroid and cause endocrine disruption. This can damage a range of organs in the body including the brain and can affect the development of children. Asthma and behavioural problems in children have increased massively in recent decades and correlations with chemicals can be considered important.

Endocrine-disrupting chemicals (EDCs) are chemicals that may mimic, block or interfere with the body's hormones, which are part of the endocrine system. Endocrine glands, distributed throughout the body, produce the hormones that act as signalling molecules after release into the circulatory system. The human body is dependent on hormones for a healthy endocrine system, which controls many biological processes like normal growth, fertility and reproduction. Hormones act in extremely small amounts, and minor disruptions in those levels may cause significant developmental and biological effects. **(51)**

A leading authority on this subject is Barbara Demeneix, whose books provide a great deal of disturbing evidence of the health effects of chemicals. **(52)**, **(53)**

Demeneix has found that chemicals in our foods, drinks, clothes and sprays are to blame for reduced IQ levels in children and higher rates of Autism Spectrum Disorder (ASD), discussed in an interview with FRANCE 24's Mairead Dundas. **(54)**, **(55)**

Chemours, a major chemical company, has taken the EU Chemicals Agency (ECHA) to court over the inclusion of GenX chemicals as a substance of very high concern (SVHC) under REACH. They lost the case.

> Endocrine-disrupting chemicals (EDCs) can cost society as a result of increases in disease and disability but—unlike other toxicant classes such as carcinogens—have yet to be codified into regulations as a hazard category. This Series paper examines economic, regulatory, and policy approaches to limit human EDC exposures and describes potential improvements. In the EU, general principles for EDCs call for minimisation of human exposure, identification as substances of very high concern, and ban on use in pesticides. In the USA, screening and testing programmes are focused on oestrogenic EDCs exclusively, and regulation is strictly risk-based. Minimisation of human exposure is unlikely without a clear overarching definition for EDCs and relevant pre-marketing test requirements. We call for a multifaceted international programme (e.g., modelled on the International Agency for Research in Cancer) to address the effects of EDCs on human health—an approach that would proactively identify hazards for subsequent regulation. **(56)**

Demeneix and her colleagues have drawn attention to the significant rise in autism. For instance, one in 250 were diagnosed in the USA in 2001, but this has risen to 1 in 68 in 2016. Autism they claim can result from a lack of thyroid hormone, which is essential for brain development in the womb or early childhood and many chemicals in buildings including flame retardants, plasticisers and other chemicals can be affecting this. Autism is a difficult subject in the UK due to the discrediting of Andrew Wakefield in 2010, who linked autism with the MMR vaccine. **(57)**

Nevertheless, there is significant scientific work which makes clear the dangers of a wide range of EDCs in buildings, not just with respiratory illnesses, asthma and cancer, but also to mental and developmental issues.

Sarah Mackenzie Ross, a clinical psychologist, has also considered whether exposure to chemical substances could be damaging your brain. When she spoke at a conference in London in 2018, she said that our capacity to produce new products outstrips our capacity to research their health effects. The rise in developmental and degenerative conditions may be linked to toxic chemicals.

> Environmental and industrial chemicals could be having an impact on psychological well-being. Should healthcare professionals be receiving training in toxicology? We live in a society awash with chemicals, some of the most common being petrol, diesel, carbon monoxide emissions, pesticide residues, plastics, flame retardants, solvents, diverse synthetic chemicals, polychlorinated biphenyls (PCBs) and metals. We can be exposed to these chemicals …, in different settings such as homes, schools or in the workplace. The manufacture and use of industrial chemicals has increased more than 15-fold in the past 70 years, and concern is growing about the impact on human health. It has been estimated that over 100,000 toxic substances are in commercial use and approximately 2300 new chemicals are developed and submitted for registration every year. **(58)**

McKenzie Ross has also carried out research in the effects of organo-phosphates and also chemical emissions in aeroplanes. Poor air quality in aircraft air conditioning has become a serious concern for the health of passengers but more significantly for the health of air crew. **(59)**

Lead

Lead is still a hazardous emission problem in buildings. While new buildings should not contain lead, a large proportion of old buildings still do. Lead pipes and lead paint are the two main issues. While lead pipes should have been replaced for the mains water supply, connections into houses and school may well still have hidden lead pipes that will contaminate drinking water. It's possible to buy lead testing kits mainly for paint and surfaces as well as water and anyone engaging painters and decorators should insist that they carry out a test.

There is an assumption that lead poisoning is uncommon and often undiagnosed by doctors, but it is more common than might be realised. The effects of lead poisoning are low birth weight and reduced growth in babies, behavioural problems in children and problems in development and education. Lead can also affect hearing. Lead can affect pregnant women leading to miscarriages, reduced foetal growth and preterm birth. In adults, lead can lead to kidney disease, increased blood pressure, depression and anxiety attacks and cardiovascular diseases. Other research has shown links to autism and possibly dementia and Alzheimer's. Despite the phasing out of lead in petrol around 20% of people can have raised levels of lead in their blood.

A project that analysed dust in homes found that 11% of UK homes have high lead levels in dust. The source of household dust is largely from renovation works where old paint has been sanded off. Both builders and DIYers rarely consider there might be a risk from lead when carrying out renovations. **(60)**

There are numerous disturbing stories on the LEAPP Alliance website on the News Personal quotes link. For instance:

> When I discovered that the renovation work myself and my 28-year-old son had done on our stairs, removing the paint over a couple of weeks, could be extremely harmful, and had most

> likely contaminated our entire house because of our ignorance of the spread of sanded paint, I had a reaction that immediately had a huge impact on my mental health. My anxiety over the situation and my guilt and fear that I had possibly harmed my son, was the worst terror I had ever experienced. I quickly I began to show severe OCD tendencies. I knew I needed to clean the contamination but my mental health had me absolutely frozen, unable to move forward. I could not have my son in the house and asked him to move out, so I am now, two months on, alone and fighting a very severe state of poor mental health, whilst trying to clean a contaminated house to make it safe to live in. **(61)**

Another example from LEAPP is the example of a family who renovated an Edwardian pub into a house. Their building contractors dry sanded old paint and contaminated the whole building with lead; even after cleaning, levels of 1500 µg/sq. ft. (10 µg/sq. ft. is USA clearance level) were found. They have not been able to move into the building. Another shocking case, currently sub-judice, is of a mother who has been charged with poisoning her son due to high lead in his system. There are many cases reported in the press of lead poisoning. **(62)**

> Most American children are exposed to lead, a substance that is not safe at any level," said co-author Dr Harvey Kaufman, a senior medical director at Quest Diagnostics, which led the study. According to the CDC, [the US Centre for Disease Control] "even low levels of lead in blood have been shown to negatively affect a child's intelligence, ability to pay attention, and academic achievement. **(63)**

Awareness of the dangers of lead contamination in the UK is not high and like so many other chemical emissions in buildings issue is not adequately addressed by authorities. The last guidance issued by the UK Government was in 2013, but a much more recent report claims deaths from lead exposure in children to be very infrequent in England but serious concerns that there is a problem is reflected in this report. **(64)**, **(65)**, **(66)**

Asbestos

Asbestos should be a problem of the past but deaths from asbestos continue to rise. Its wise not to take statements in the Daily Mail too seriously, but there is evidence that many pupils and teachers have died as a result of asbestos since 1980. **(67)**

The Joint Union Asbestos Committee (JUAC) has claimed that 17 schoolteachers died of mesothelioma (a cancer linked to asbestos exposure) in 2017. It is a legal requirement to carry out asbestos checks in buildings and normally a requirement for it to be safely removed by approved contractors. However, the Health and Safety Executive takes the view that "wherever possible asbestos containing materials should be left in situ, managed and in some cases encapsulated."

The Joint Union Asbestos Committee (JUAC) claim that the UK Government has no policy of phased removal of asbestos. **(68)**

The Labour Research Department states that asbestos is still present in at least 451 NHS premises and buildings in London alone, and 695 in Scotland. Moreover, 31 local authorities in England own 2,690 buildings with asbestos (not including schools and housing). **(69)**

The National Education Union is particularly concerned about this problem due to high levels in schools, not just in pipe insulations but there have been insulation boards used as pin boards in classrooms and asbestos dust is released when drawing pins and staples are put into these boards. **(70)**

There was surprisingly little academic research on asbestos in buildings available through a literature search. Work published by Gilham et al in 2018 unsurprisingly calls for more research, having stated that the risk of asbestos exposure in "Britain" is about 1 in 10,000, but they make the surprising claim that risks from asbestos in older buildings are not known and yet:

> Our results suggest that a minority of the general population may have unusually high environmental exposure, but more sensitive fibre counting will be needed to confirm this. **(71)**

Asbestos was used in backing for cheap vinyl tiles, used in their millions in social housing and asbestos contamination has been detected in houses that are now being retrofitted. Many walls and ceilings were coated with a product called Artex and up to a certain date this could have been made of asbestos. Often this is sanded off, leading to asbestos contamination, or sometimes it is covered over with a new plaster. There are numerous websites providing a range of good and bad advice about how to deal with Artex and whether it contains asbestos. **(72)**, **(73)**

There is a useful paper on asbestos in 1970s buildings in Coventry, when they were being renovated. Asbestos was found in fixing holes to concrete around windows which is a clear example of the myriad of uses of asbestos that might be in unexpected places. **(74)**

Trade unions have been active around this issue as the risks to construction workers and many other employees is much higher than the general population. The London Hazards centre has a Mesothelioma day, the last of which was held on July 7 2023. They draw attention to external materials, particularly in roof construction, that have incorporated asbestos. **(75)**

Eternit is a well-known name for construction products and at one time made products which contained asbestos. An Eternit company went bankrupt in 1986 but Eternit still exists as a business in the UK, also operating under the name Marley. Stephan Schmidheiny, a former major shareholder in Eternit, was found guilty of causing the deaths of 392 people in Casale Monferrato in Piedmont. Schmidheiny was told to pay €50 million in provisional damages, and €500 million to an association looking after the relatives of victims of asbestos. **(76)** The Schmidheiny family are Swiss billionaires with interests in the Lafarge Holcim cement giant. **(77)** Eternit asbestos and fibre cement products were also made in Belgium and the UK. **(78)**

The Health and Safety Executive produces a very useful list of products that might contain asbestos, nearly a hundred entries long, though they warn that it might not be exhaustive. **(79)**

References

(1) Weschler, C. J. (2009). 'Changes in indoor pollutants since the 1950s,' *Atmospheric Environment*, 43(1), 156–169. Available at: https://doi.org/10.1016/j.atmosenv.2008.09.044 [Accessed 17 Sep. 2023].

(2) ECHA (n.d). *REACH – Registration, Evaluation, Authorisation and Restriction of Chemicals Regulation.* Available at: https://echa.europa.eu/information-on-chemicals/registered-substances [Accessed 17 Sep. 2023]

(3) Watson, A. (2021). 'Why it's time for a bold UK Chemicals Strategy.' *Chemtrust.* Available at: https://chemtrust.org/bold-uk-chemicals-strategy/ [Accessed 17 Sep. 2023].

(4) Weschler (2009). Op cit.

(5) Crump, D. et al. (2016) 'Health-based evaluation of chemical emissions from construction products to indoor: Development and application of the EU-LCI harmonisation framework.' *Proceedings of Indoor Air.* 3–8 July. Ghent, Belgium, paper 96.

(6) IAQ Consulting (n.d). *Progress with the establishment of harmonised health-based reference values, EU-LCIs.* [pdf]. Available at: www.umweltbundesamt.de/sites/default/files/medien/421/dokumente/06_crump_eu_lci.pdf [Accessed 17 Sep. 2023].

(7) Available at: https://single-market-economy.ec.europa.eu/sectors/construction/eu-lci-subgroup/eu-lci-values_en

(8) Available at: https://ec.europa.eu/docsroom/documents/49239

(9) Morantes, G., Jones, B., Molina, C. and Sherman, M. H. (2023). 'Harm from indoor air contaminants.' *SSRN*. Available at https://ssrn.com/abstract=4409736 [Accessed 17 Sep. 2023].

(10) Holøs, S. B., Yang, A., Lind, M. Thunshelle, K., Schild, P. and Mysen, M. (2019). 'VOC emission rates in newly built and renovated buildings, and the influence of ventilation – a review and meta-analysis,' *International Journal of Ventilation*, 18(3), 153–166. Available at: https://doi.org/10.1080/14733315.2018.1435026 [Accessed 17 Sep. 2023].

(11) Huang, L., Anastas, N., Egeghy, P., Vallero, D. A., Jolliet, O. and Bare, J. (2019). 'Integrating exposure to chemicals in building materials during use stage', *The International Journal of Life Cycle Assessment*, 24(6). Available at: https://doi.org/10.1007/s11367-018-1551-8 [Accessed 17 Sep. 2023].

(12) Jolliet, O., Ernstoff, A. S., Csiszar, S. A. and Fantke, P. (2015). 'Defining product intake fraction to quantify and compare exposure to consumer products.' *Environmental Science Technology*, 49, 8924–8931. Available at: https://doi.org/10.1021/acs.est.5b01083 [Accessed 17 Sep. 2023].

(13) Jolliet, O., Huang, L., Hou, P. and Fantke, P. (2021). 'High throughput risk and impact screening of chemicals in consumer products.' *Risk Analysis: An Official Publication of The Society For Risk Analysis*, 41(4), 627–644. Available at: https://doi.org/10.1111/risa.13604 [Accessed 17 Sep. 2023].

(14) Berge, B., Butters, C., and Henley, F. (2009). *The Ecology of Building Materials*. London: Routledge.

(15) Email conversation with T. Roberts, HPD Collaborative USA. Jan 30, 2022.

(16) Carter, T. et al. (2023). 'A modelling study of indoor air chemistry: The surface interactions of ozone and hydrogen peroxide,' *Atmospheric Environment*, 297, 119598. Available at: https://doi.org/10.1016/j.atmosenv.2023.119598 [Accessed 17 Sep. 2023].

(17) USETOX. (2023). USEtox Corrective Release 2.13. Available at: www.usetox.org/model/download/usetox2.13 [Accessed 17 Sep. 2023].

(18) PHAROS. (n.d). Search Pharos. Available at: https://pharosproject.net/ [Accessed 17 Sep. 2023].

(19) HBN. (n.d). Product Guidance. Available at: https://healthybuilding.net/products [Accessed 17 Sep. 2023].

(20) The use of the term LCI should not be confused with the LCI database referred to above and one has to wonder who failed to spot the potential for confusion between LCI and LCI. Was this EU bureaucratic stupidity or deliberately meant to confuse?

(21) Huang, L., Fantke, P., and Jolliet, O. (2021). 'consumer products as a source of human exposure to chemicals.' *Environmental Pollutant Exposures and Public Health.* 50 edn, vol. 2021, Issues in Environmental Science and Technology, pp. 295–352. Available at: https://doi.org/10.1039/9781839160431-00295 [Accessed 17 Sep. 2023].

(22) Huang, et al. (2021). Op cit.

(23) Fantke, P., & Jolliet, O. (2016) 'Life cycle human health impacts of 875 pesticides.' *International Journal of Life Cycle Assessment*, 21(5), Article 722–733. Available at: https://doi.org/10.1007/s11367-015-0910-y

(24) Morantes, G., Jones, B., Molina, C. and Sherman, M. H. (2023). 'Harm from indoor air contaminants.' *SSRN* Available at: https://ssrn.com/abstract=4409736 or http://dx.doi.org/10.2139/ssrn.4409736 [Accessed 17 Sep. 2023].

(25) Email from Benjamin Jones Nottingham University April 24, 2023.

(26) Huang et al. (2021). Op cit.

(27) Maury-Micolier, A., Huang, L., Taillandier, F., Sonnemann, G. and Jolliet, O. (2023) 'A life cycle approach to indoor air quality in designing sustainable buildings: Human health impacts of three inner and outer insulations.' *Building and Environment*, 230. Available at: https://doi.org/10.1016/j.buildenv.2023.109994 [Accessed 17 Sep. 2023].

(28) PU Europe. (n.d). Available at: www.pu-europe.eu/ [Accessed 17 Sep. 2023].

(29) Zucco, G. M. and Doty, R. L. (2022). 'Multiple chemical sensitivity.' *Brain Science*, 2022(1), 46. Available at: www.ncbi.nlm.nih.gov/pmc/articles/PMC8773480/#:~:text=Multiple%20Chemical%20Sensitivity%20(MCS)%20is,%2C%20epoxy%2C%20pesticides%2C%20and%20some [Accessed 3 October 23].

(30) Torii, S. (2000). 'Sick House syndrome, sick building syndrome, indoor harmful substance sensitivity.' *PubMed.* 31, 609–12. Available at: https://pubmed.ncbi.nlm.nih.gov/11269178/ [Accessed 17 Sep. 2023].
(31) Nordin, S. (2020). Mechanisms underlying nontoxic indoor air health problems: A review. *International Journal of Hygiene and Environmental Health*, 226, 113489. Available at: https://doi.org/10.1016/j.ijheh.2020.113489 [Accessed 17 Sep. 2023].
(32) Hoffmann, H. D., and Schupp, T. (2009). 'Evaluation of consumer risk resulting from exposure against diphenylmethane-4,4'-diisocyanate (MDI) from polyurethane foam.' *EXCLI Journal*, 8, 58–65. Available at: https://doi.org/10.17877/DE290R-579 [Accessed 17 Sep. 2023].
(33) Maury Micollier (2023). Op cit.
(34) Mawditt, I. (2017). Don't play with radon. Available at: www.fourwalls-uk.com/blog/ [Accessed 17 Sep. 2023].
(35) Maury Micollier (2023). Op cit.
(36) He, Z. et al. (2019). 'An improved mechanism-based model for predicting the long-term formaldehyde emissions from composite wood products with exposed edges and seams.' *Environment International*, 132, 105086. Available at: https://doi.org/10.1016/j.envint.2019.105086 [Accessed 17 Sep. 2023].
(37) Maury Micollier (2023). Op cit.
(38) Pelletier, M. et al. (2017). 'Dermal absorption of semi-volatile organic compounds from the gas phase: Sensitivity of exposure assessment by steady-state modelling to key parameters.' *Environment International*, 102, 106–113. Available at: https://doi.org/10.1016/j.envint.2017.02.005 [Accessed 17 Sep. 2023].
(39) Perkins, T. (2023). 'Societal cost of 'forever chemicals' about $17.5tn across global economy-report.' *The Guardian.* Available at: www.theguardian.com/environment/2023/may/12/pfas-forever-chemicals-societal-cost-new-report# [Accessed 17 Sep. 2023].
(40) CHEMSEC. (n.d). PFAS Movement. Available at: https://chemsec.org/pfas/ [Accessed 17 Sep. 2023].
(41) Morales-McDevitt, M. E., Becanova, J., Blum, A., Bruton, T. A., Vojta, S., Woodward, M. and Lohmann, R. (2021). 'The Air that We Breathe: Neutral and volatile PFAS in Indoor Air.' *Environmental science & technology letters*, 8(10), 897–902. Available at: https://doi.org/10.1021/acs.estlett.1c00481 [Accessed 17 Sep. 2023].
(42) Trager, R. (2021). 'Efforts underway in Europe to ban PFAS compounds.' *Royal Society of Chemistry*. Available at: www.chemistryworld.com/news/efforts-underway-in-europe-to-ban-pfas-compounds/4014038.article [Accessed 17 Sep. 2023].
(43) Fernandez, S. R., Kwiatkowski, C., and Bruton, T. (2021). 'Building a better world: Eliminating Unnecessary PFAS in building materials.' *Green Science Policy Institute*. Available at: https://greensciencepolicy.org/docs/pfas-building-materials-2021.pdf [Accessed 17 Sep. 2023].
(44) Available at: www.euronews.com/green/2023/08/04/forever-chemicals-and-blood-cancer-jersey-residents-consider-class-action-lawsuit-over-pfa [Accessed 19 Sep. 2023].
(45) Owen, M. (2023). 'Jersey "forever chemicals" contaminated private water supply.' *BBC NEWS.* Available at: www.bbc.co.uk/news/world-europe-jersey-66361242 [Accessed 17 Sep. 2023].
(46) Marales-McDevitt, M. E., Becanova, J., Blum, A., Bruton, T. A., Vojta, S., Woodward, M. and Lohmann, R. (2021). 'The air that we breathe: Neutral and volatile PFAS in indoor air.' *Environmental Science & Technology Letters,* 8(10), 897–902.
(47) Cahuas, L., Muensterman, D., J., Kim-Fu, M. L., Reardon, P. N., Titaley, I. A. and Field, J. A. (2022). 'Paints: A source of volatile PFAS in air—potential implications for inhalation exposure.' *Environmental Science & Technology*, 56(23), 17070–17079. Available at: https://doi.org/10.1021/acs.est.2c04864 [Accessed 17 Sep. 2023].
(48) Perkin, T. (2021). 'Toxic 'forever chemicals' contaminate indoor air at worrying levels, study finds.' *The Guardian.* Available at: www.theguardian.com/society/2021/aug/31/pfas-toxic-forever-chemicals-air-breathing [Accessed 17 Sep. 2023].

(49) VECTOR FOILTEC. (n.d). Eden Project. Available at: www.vector-foiltec.com/projects/eden-project-botanic-sustainable/ [Accessed 17 Sep. 2023].

(50) Marales-McDevitt (2021). Op cit.

(51) NIH (2023). Endocrine Disruptors. Available at: www.niehs.nih.gov/health/topics/agents/endocrine/index.cfm [Accessed 17 Sep. 2023].

(52) Demeneix, B. (2017). *Toxic Cocktail: How Chemical Pollution is Poisoning Our Brains*. New York, New York: Oxford University Press.

(53) Demeneix, B. (2014) 'How environmental pollution impairs human intelligence and mental health' *Oxford Series in Behavioural Neuroendocrinology*.

(54) Dundas, M. (2017). 'Chemicals in our everyday environment "are poisoning our brains."' *France 24*. Available at: www.france24.com/en/20171223-interview-barbara-demeneix-chemicals-toxic-cocktail-poisoning-brains-iq-autism-endocrines [Accessed 17 Sep. 2023].

(55) Demeneix, B. (2022). 'Fighting endocrine disruption: Are we getting somewhere?' *Endocrine Abstracts,* 81. Presented at European Congress of Endocrinology 2022, Milan, Italy. Available at: www.endocrine-abstracts.org/ea/0081/ea0081ecas1.6 [Accessed 17 Sep. 2023].

(56) Kassotis, C. D. (2020). 'Endocrine-disrupting chemicals: economic, regulatory, and policy implications.' *The Lancet Diabetes & Endocrinology*, 8(8), 719–730. https://doi.org/10.1016/s2213-8587(20)30128-5 [Accessed 17 Sep. 2023].

(57) Sathyanarayana Rao, T. S. and Andrade, C. (2011). 'The MMR vaccine and autism: Sensation, refutation, retraction, and fraud.' *Indian Journal of Psychiatry*, 53(2), 95. https://doi.org/10.4103/0019-5545.82529 [Accessed 17 Sep. 2023].

(58) Ross, S. M. (2017). 'Hazardous to health?' *The Psychologist.* Available at: https://issuu.com/thepsychologist/docs/psy1017big_ccd505d7dc5a88 [Accessed 17 Sep. 2023].

(59) Ross, S. M. (2013). 'Neurobehavioral problems following low-level exposure to organophosphate pesticides: A systematic and meta-analytic review.' *Critical Reviews in Toxicology*, 43(1), 21–44. Available at: https://doi.org/10.3109/10408444.2012.738645 [Accessed 17 Sep. 2023].

(60) Pye, T. (2021). 'Toxic lead still here, still harming: A manifesto for a lead-safe UK.' *Lead Safe World* [pdf]. Available at: https://leadsafeworld.com/wp-content/uploads/2021/04/A-Manifesto-for-a-Lead-Safe-UK-d0.6.pdf [Accessed 17 Sep. 2023].

(61) Pye, T. (n.d). LEAPP Alliance website. Available at: https://leappalliance.org.uk/ [Accessed 17 Sep. 2023].

(62) McCormick, E., and Lutz, E. (2021). "We're losing IQ points': the lead poisoning crisis unfolding among US children." *The Guardian.* Available at: www.theguardian.com/us-news/2021/dec/08/lead-poisoning-crisis-us-children [Accessed 17 Sep. 2023].

(63) CDC. (2022). Health Effects of Lead Exposure. Available at: https://www.cdc.gov/nceh/lead/prevention/health-effects.htm [Accessed 17 Sep. 2023].

(64) Lead Exposure in Children Surveillance System (LEICSS) Annual Report 2022 Summary of 2021 Data Health Protection Report Volume 17 Number 1 12 January 2023.

(65) Department for Environment, Food & Rural Affairs. (2013). Advice on lead paint in older homes. Available at: www.gov.uk/government/publications/advice-on-lead-paint-in-older-homes [Accessed 17 Sep. 2023].

(66) UK Health Security Agency. (2023). Lead Exposure in Children Surveillance System (LEICSS) annual report 2022. [pdf] 17(1) Available at: https://assets.publishing.service.gov.uk/government/uploads/system/uploads/attachment_data/file/1128326/hpr0123_LEICSS_2021.pdf [Accessed 17 Sep. 2023].

(67) Woodland, D. (2023). 'Classroom asbestos kills 10,000 pupils and teachers in four decades – as fears raised 21,000 UK schools are STILL riddled with deadly toxic fibres.' *Daily Mail.* Available at: www.dailymail.co.uk/news/article-12255119/Classroom-asbestos-kills-10-000-pupils-teachers-four-decades.html [Accessed 17 Sep. 2023]

(68) JUAC. (n.d). Key Facts. Available at: https://the-juac.co.uk/school-college-facts/ [Accessed 17 Sep. 2023].

(69) Labour Research Department. (2023). Asbestos in public buildings. Available at: www.lrd.org.uk/research/asbestos-public-buildings [Accessed 17 Sep. 2023].

(70) National Education Union. (2023). Asbestos in schools. Available at: https://neu.org.uk/advice/health-and-safety/work-environment/asbestos-schools [Accessed 17 Sep. 2023]
(71) Gilham, C., Rake, C., Hodgson, J., Darnton, A., Burdett, G. and Wild, J. P. (2018). 'Past and current asbestos exposure and future mesothelioma risks in Britain: The Inhaled Particles Study (TIPS).' *International Journal of Epidemiology*, 47(6), 1745–1756. https://doi.org/10.1093/ije/dyx276 [Accessed 17 Sep. 2023].
(72) BEACON. (2021). How to remove Artex coatings. Available at: www.beaconinternational.co.uk/how-to-remove-artex-coatings/#:~:text=Is%20it%20dangerous%20to%20scrape,can%20disrupt%20the%20hazardous%20fibres [Accessed 17 Sep. 2023].
(73) Crucial Environmental. (2022). Is Artex hiding danger in your home? Available at: www.crucial-enviro.co.uk/blog/is-artex-hiding-danger-in-your-home/ [Accessed 17 Sep. 2023].
(74) Mateo-Garcia, M., Ahmed, A. and Mcdough, D. (2017). Non-invasive approaches for low-energy retrofit of buildings: Implementation, monitoring and simulation in a living lab case study *Structural Studies, Repairs and Maintenance of Heritage Architecture XV*. Available at: https://doi.org/10.2495/str170161. [Accessed 17 Sep. 2023].
(75) London Hazards Magazine Action Mesothelioma Action Day. (2023). Available at: www.lhc.org.uk/london-hazards-magazine-action-mesothelioma-action-day-2023/ [Accessed 17 Sep. 2023]
(76) Giuffrida, A. (2023). 'Swiss billionaire jailed over asbestos-related deaths in Italian town.' *The Guardian*. Available at: www.theguardian.com/world/2023/jun/08/stephan-schmidheiny-swiss-billionaire-jailed-over-asbestos-related-deaths-piedmont-italy [Accessed 17 Sep. 2023].
(77) Forbes. (2023). *Thomas Schmidheiny*. Available at: www.forbes.com/profile/thomas-schmidheiny/?sh=2c4495d41e6f [Accessed 17 Sep. 2023].
(78) Wikipedia. (2023). Eternit. Available at: https://en.wikipedia.org/wiki/Eternit [Accessed 17 Sep. 2023].
(79) HSE. (n.d). Products that might contain asbestos. Available at: www.hse.gov.uk/asbestos/managing/products.htm [Accessed 17 Sep. 2023].

4 UK, EU and WHO policies on indoor air quality

The general population are meant to be protected from dangers and hazards by Government legislation. The Health and Safety Executive, Environmental Health and the Housing Health and Rating System (HHSRS), building and fire regulations and so on exist to protect us all in our homes and workplaces, but despite being a wealthy country, the UK continues to have some of the worst health and poverty problems in Europe. As will be explored in Chapter 5, many people face serious problems from mould and damp in their homes, an issue going back to the 19th century. Less well-off and vulnerable people often face bad housing conditions, often scared to complain as they fear homelessness. The UK is said to face a shortage of 4 million homes with 18% of young people living in poor quality housing. The scale of the problem came to the fore following the Grenfell tower fire on June 14, 2017. Not only was this an appalling disaster with 72 deaths and lives of those who escaped and lived nearby traumatised, as Hodkinson explains, it brought into the open a much wider issue about working class housing in the UK. **(1)**

It was decided that the Grenfell Inquiry would not include discussion of wider issues relating to social housing policy in its terms of reference, though anyone following the Grenfell Inquiry hearings cannot avoid drawing conclusions about the attitudes, almost verging on contempt, of Government, local authorities, professionals and construction material suppliers, toward less advantaged people. **(2)**

The Grenfell Inquiry demonstrated how companies, anxious to gain market share for their products, were happy to lie and cheat and game the rules. As Apps explains in his book about Grenfell and in his Chapter 10 titled "We will be Quids in"

> It is now clear how dangerous cladding and insulation materials became widely available in the UK: the bitter fruit of corporations, capitalising on deregulation … . **(3)**

Apps gives a brief glimpse into a murky world in which business interests have been given far too much influence in setting standards and regulations, with their main aim being to ensure that they continue to sell their products. Many examples can be given of the scale of influence of industry on regulations. For instance, a British Standards Institute working group on sustainability in construction is largely made up of vested interest construction materials companies. Of a meeting of less than 30 people, 21 are from construction industry interest groups and even some of the independent individuals may represent certain construction groups.

Many of the people who also contribute to discussions of regulations in European standards bodies, the International Standards Organisation representing the UK, do little to disguise their lobbying on behalf of concrete blocks, cement, steel and petrochemical and plastics materials.

DOI: 10.1201/9781003129226-4

Table 4.1 BSI 558 committee members

BSI 558 committee concerned with sustainability has only three or four members independent of construction vested interests
Here are examples of the organisations normally represented at meetings:
Precast concrete division of the Mineral Products Association
Mineral Products Association
Building Research Establishment x 3
Concrete Block Association
Wood Panel Industries Federation
European Resilient Floor Manufacturers Institute
Civil Engineering Award Scheme
Council for Aluminium in Building
Insulation Manufacturers Association (promoting PIR insulation)
British Constructional Steelwork Association
UK Steel
Atkins Engineering Consultancy
United Kingdom Quality Ash Association
Institution of Structural Engineers
Mineral Wool Manufacturers Association
Lucideon Materials testing
British Woodworking Federation
Construction Products Association
British Association of Reinforcement
Cement Admixtures Association
Galvanizers Association
Royal Institution of Chartered Surveyors
HS2 London Railway Tunnels
British Constructional Steelwork Association

BSI committees are linked to EU standards committees and working groups, and will comment on proposed changes to EU policies, even though the UK is no longer in the EU. CEN/TC350 is the overall committee responsible for EU standards on sustainable construction. In the summer of 2023 it was proposed that Working Group 1 of CEN/TC350 should deal with the issue of emissions to indoor air. This was an important step forward to get indoor air quality and emissions from building materials on the agenda and the BS committee was given the opportunity to comment (known as a ballot). However, out of 31 members only three bothered to comment and only two in favour of this proposal. This is a clear indication of the lack of interest in indoor air pollution on the part of mainstream industry "experts" on sustainable construction.

"Independent" expert advisors

When Government brings in experts and advisory bodies to provide guidance, such as following the Grenfell disaster, they often turn to people who have clear conflicts of interest. For instance, Dame Judith Hackitt who was appointed to chair an "independent" review of the building regulations and fire safety, following the Grenfell disaster, with her proposals forming the basis of some of the Government's subsequent legislation, had a clear relationship with the chemicals industry.

> It would be difficult to find someone who is less independent than Judith Hackitt. She is a complete chemical industry insider. She was also a commissioner of the useless Health and Safety Executive. She is on the Board of the Energy Saving Trust which actively promote toxic foam insulation. **(4)**

Hackitt is possibly no longer involved with the Energy Saving Trust and HSE but was a director at the Chemical Industries Association from 1998 to 2005 and also worked for the European Chemical Industries Council to 2007, during which time industry may not have done enough to ensure the safety of petrochemical insulation products. They worked against the tightening of EU regulations according to the *Financial Times* (2021):

> Europe's chemicals industry has warned that EU efforts to tighten the regulation of more than 10,000 substances threaten its future just as it grapples with the pressures to reduce its greenhouse gas emissions. **(5)**

Hackitt was a director of the Energy Saving Trust (EST) which used to provide explicit support for the use of petrochemical foam insulation products on its website. However, the EST appears to have withdrawn its endorsement of products, though support for some hazardous technologies are still implied in its publication of "case studies." **(6), (7), (8)**

The Government also appointed Sir Ken Knight chair of the Building Safety "independent" (sic) expert advisory panel in 2017 following Grenfell. **(9)**

Knight had been an advisor on Fire and Rescue between 2007 and 2013 having been Commissioner of the London Fire Brigade from 2003 to 2007 and was also a director of Warrington-fire, the private fire testing and certification firm, between 2004 and 2017. As part of his work there, and according to the Fire Brigades Union (FBU), he certified similar cladding to that fitted on Grenfell Tower. According to the FBU, Warrington fire shared a parent company with the fire safety consultant for the refurbishment of Grenfell. Matt Wrack, general secretary of the FBU, commented:

> In the years before the Grenfell Tower tragedy, Sir Ken Knight was a senior advisor to the Government on the subject of fire safety. He needs to accept his share of culpability. In the run-up to the fire, Government ministers took an axe to the UK's Fire and Rescue Service and fire safety regulation in general, and Sir Ken Knight helped provide cover for them to do that. He would also have had countless opportunities to raise concerns around key issues that would later contribute to Grenfell, but he failed to do so. **(10), (11), (12)**

The Grenfell Inquiry report, when it eventually emerges in 2024, will be a massive document, and for those who choose to read it will be a devastating indictment of the failure of the regulatory system to protect the health and well-being of the occupants of buildings. New laws have been passed which include a Building Safety Act and a Fire Safety Act. Various disparate bits of the jigsaw have been absorbed into the already overstretched Health and Safety Executive including a new figure known as the Building Safety Regulator. **(13)**

Many of the measures introduced appear to establish a stricter regulatory regime but the fundamental flaws of the system exposed by the Grenfell Inquiry may not have been adequately addressed.

The Health and Safety Executive has a responsibility for maintaining adequate ventilation in workplaces to safeguard workers, but does not have overall responsibility for ensuring good indoor air quality. Indoor air quality has received little attention by Government and regulatory bodies, as they have been largely pre-occupied with traffic pollution.

Masking the issue with a focus on ambient air pollution

The focus of the UK Government has been mainly on ambient/outdoor air quality and emissions from traffic, industrial and agricultural pollution. When investigating the UK literature on air pollution, it would be easy to miss indoor air quality entirely as the focus on traffic and external air pollution is all pervasive. A number of scientific, public sector and non-government organisations (NGOs) even refuse to consider indoor air quality at all or pay limited lip service to the topic.

The UK Government Air Quality Framework has been largely driven by EU and WHO standards on traffic pollution, which the UK has failed to meet on a few occasions. Efforts to reduce traffic pollution have led to policies encouraging electric cars and low-emissions zones in some towns and cities. There is some mention of Volatile Organic Compounds (VOCs) in the Government framework, but it does not refer to the word *indoor* once in its 53 pages, despite claiming that it is promoting policies and regulations to reduce harmful emissions that can damage human health. **(14)**

On the other hand, the UK Government Ministry Department of Environment, Food and Rural Affairs, (DEFRA) has begun to recognise the problem of indoor air quality through its Air Quality Expert Group (AQEG) and produced a report about the issue in 2022, but this document tends to play down the risks to health from building materials emissions. It does recognise a problem of off-gassing from formaldehyde and VOCs but refers to these in the context of what they call "everyday products." Off-gassing is only mentioned once in 142 pages though the report does list European labelling systems that touch on building materials emissions. **(15)** Flame retardants, phthalates and SVOCs get a brief mention in the DEFRA AQEG report, but are linked to emissions from combustion products. Building materials are discussed in the context of moisture but the general view seems to be that "typical indoor sources (are) skin oils, cooking and cleaning." (p. 83). Indeed, building materials are only mentioned **nine** times in the whole report.

There is some rather confusing discussion in the DEFRA report of "partitioning indoor materials" citing work by Algrim et al but without a satisfactory explanation of what this means:

Table 4.2 Different guidance on products in EU countries

Table 2.3: Building materials, product labels on chemical emissions in EU (from DFE, 2018. BB101 Guidance on ventilation, thermal comfort and indoor air quality in schools) Building materials and products labels and guidance on chemical emissions in EU

- European Ecolabel (e.g., textile-covered flooring, wooden flooring, mattresses, indoor and outdoor paints and varnishes: Europe) **(16)**
- EMICODE® (adhesives, sealants, parquet varnishes and other construction products: Germany/Europe) **(17)**
- GUT (carpets: Germany/Europe) **(18)**
- Blue Angel (Germany) **(19)**
- Nordic Swan (Scandinavia), https://www.nordic-swan-ecolabel.org/
- Umweltzeichen (Austria) **(20)**
- AgBB (Specifications for construction products: Germany) **(21)**
- M1 (construction products: Finland) **(22)**
- ANSES (formerly AFSSET) (construction products: France) **(23)** https://www.anses.fr/en
- CertiPUR (PU foam for furniture industry: Europe) **(24)**
- Danish Indoor Climate Label **(25)** https://www.ecolabelindex.com/ecolabel/danish-indoor-climate-label
- Swedish 'byggvarudeklaration' **(26)** https://www.byggvarubedomningen.se/
- Natureplus (construction products: Germany/Europe **(27)**

This list was derived from the AQEG Indoor Air Quality Report DEFRA 2022 **(15 op cit)** that contained numerous errors, which have been corrected as far as possible.

Attention is also now being given to **partitioning** to indoor materials themselves, such as paintwork and upholstery, which dominate the available partitioning volume
of absorptive partitioning of volatile organic compounds to painted surfaces.

Algrim explains that 50% of VOCs indoors in his research were octane, hexanone and propanol from paint. **(28)** Given the thousands of possible citations of scientific work on emissions from building materials it is strange that this is the only reference included by AQEG to materials, presumably to suggest that the main indoor emissions are from paints.

By comparison ventilation receives 189 mentions, though the report does at least refer to inadequacies in mechanical and MVHR systems. The DEFRA report does go into a little detail about VOCs, including a table (1.1) which lists some emissions. **(29)**

> VOCs may be emitted over periods of months or years from a wide range of construction and finishing products that are used in a building. Examples include concrete and masonry surface treatments, timber preservation and coatings, adhesives, sealants, paints and coatings, damp-proofing emulsions and membranes, wall coverings, floor coverings and fungicide washes. In addition, the volatile organic compound (VOC) formaldehyde may be emitted over significant periods of time from resins, phenol-formaldehyde and urea formaldehyde (UF) from wood-based products such as particleboard in furniture, urea-formaldehyde based lacquers and foam cavity wall insulation. **(29)**

However, this is immediately followed by a discussion of "ranges of consumption patterns" while referring to brick built homes! This unsatisfactory list does hint that there is a problem of emissions from building materials, though the report itself almost entirely ignores the issue. The confusion caused by this document is hardly surprising as the expert group enlisted to produce it did not seem to include any experts on building material emissions apart from someone from the BRE.

It's not easy to explain why DEFRA and its scientists should be so coy about this issue, but Professor Alistair Lewis of York University, the chair of AQEG, has published several papers which gives the impression that the problem of VOCs is mainly from external air pollution and cleaning products. In a short paper, "The Changing Face of Urban Air Pollution," he discusses VOC in Urban air with very little mention of indoor emissions. He claims that VOCs from cleaning products dominate the air in modern homes. **(30)**

There is no doubt that cleaning materials, air fresheners, cosmetics and emissions from outside homes can have a bad effect on indoor air quality, but this is no excuse for ignoring emissions from building materials. In another paper, Lewis and the Chief Medical officer Chris Whitty provide only six indicators of indoor air quality, wood burning, cosmetics and toiletries, paint, natural gas combustion, social housing energy efficient rating (which is not really an indicator of indoor air quality) and damp problems (mould) in homes. Building materials (only mentioned twice) are not listed as a source of bad indoor air quality, though formaldehyde and VOCs are briefly mentioned. They do acknowledge that indoor air contains a more diverse range of pollutants than does outdoor air and state that we need better indoor air quality, but create confusion around the issue. They refer to a much greater awareness of indoor air quality in France but fail to refer to policies in other European countries.

> Building materials, fabrics and furniture also give off chemicals that can irritate the lungs and eyes. Volatile organic compounds are released from paints, carpets and wood treatments and other household products.

Local and national governments must ensure that good indoor air quality is delivered for those in shared, social or rented accommodation, and for public indoor spaces. For example, in France, monitoring of a range of pollutants is mandatory in schools. Beyond state intervention, employers must ensure safe, healthy workplaces, including good-quality air. **(31)**

The role of the Building Research Establishment (BRE)

Dr. Andy Dengel, head of indoor air quality testing at the BRE (now known as the BRE Group), helped to write the DEFRA report. The BRE building rating system BREEAM does include a detailed set of standards on indoor air quality though Dengel himself only lists two of his own publications, which are about heating and energy efficiency, and it was not possible to find any BRE research into emissions from building materials though Dengel and others have written a useful paper warning of problems related to ventilation. **(32), (33)**

> Many of the potential hazards to health associated with energy saving in homes depend upon whether or not sufficient ventilation is provided. Whilst there is evidence to link ventilation to indoor air pollutants, and indoor air pollutants to health, there is less information about the direct links. There has never been a comprehensive study on the role of home ventilation for ensuring health; of ventilation rates achieved in practice in UK homes; or a definitive assessment of a safe minimum level of ventilation (although 0.5 air changes per hour is widely recommended). It is also notable that most published studies of energy efficient homes use a limited definition of occupant health, focusing on thermal comfort and occupant satisfaction. **(34)**

The BRE tis currently advertising training on indoor air quality, but it is of a very superficial introductory nature. The BRE offers a sustainable building assessment tool called BREEAM that includes a detailed section on indoor air quality (HEA2), which is an excellent and detailed analysis of indoor air quality issues and low-emissions products, and gets a brief mention in the AQEG report. **(35),(36)**

The DEFRA AQEG report distracts attention from emissions from building materials by placing undue emphasis on emissions from cooking and wood burning stoves, and this is discussed below.

NICE: indoor air quality in the home

The National Institute for Health and Care Excellence has an important role to play in terms of health policies and approving medicines in the UK, setting quality standards and indicators and is highly respected. Thus, expectations about its recent study of indoor air quality in the home were high. NICE spent several years investigating the issue with a committee of experts. Their resulting report says that "People may be particularly vulnerable to ill health as a result of exposure to poor indoor air quality."

NICE has made recommendations for local authorities to:

> (1.1.1) "Embed a plan for improving indoor air quality in an existing strategy or plan to improve people's health. This could be a general air quality strategy if one exists. Otherwise, for example, include it in a strategy on housing, health and wellbeing, or inequalities."
>
> (1.1.5) Encourage the use of existing home visits to identify poor indoor air quality. For example, visits to people's homes by housing officers, environmental health practitioners, community health services, social workers, care workers, and fire and rescue services.

The report refers to people with asthma, other respiratory conditions or cardiovascular conditions, stating that:

> (1.5.1) Explain that indoor air pollutants (including nitrogen dioxide, damp, mould, particulate matter and VOCs) can trigger or exacerbate asthma, other respiratory conditions or cardiovascular conditions.

The report also refers to regulators and building control teams and calls for improvements in regulations, referring, however, to existing inadequate public health guidelines.

> (1.6.1) Update existing standards, for example building regulations, or develop new ones for indoor air quality. Base them on current safe limits set for pollutants in residential developments. See, for example, World Health Organization guidelines on selected pollutants (2010) and dampness and mould (2009), and the Public Health England indoor air quality guidelines for selected VOCs (2019).

The NICE report states that there are regulations on pollutant threshold levels but information on the level of emissions from different materials is limited admitting that few regulations exist to guide the choice of materials according to their effect on indoor air quality. Unlike DEFRA and AQEG they do acknowledge the importance of source control:

> Usually the most effective way to deal with indoor pollutants is to either remove the source or reduce emissions from it. If these are not possible, the pollutant can be diluted by ventilation (for example, opening windows) to reduce exposure.

They advise people to choose low-emission materials (e.g., products with a low volatile organic compound [VOC] or formaldehyde content and emissions) if furniture or flooring needs replacing. Unfortunately, NICE failed to investigate emissions from materials or to seek advice from experts in the field on this issue, and so their guidance is very weak. For instance, their guidance to architects and designers on how to avoid sources of pollutants is vague and inadequate:

> (1.7.1) Consider specifying building materials and products that only emit a low level of formaldehyde and VOCs. Use existing labelling schemes or other available information on product emissions (for example, on product labels) to make these specifications.

NICE recognises, but fails to discuss, the fact that labelling schemes and information on product emissions barely exist in the UK, so it is unlikely that they will be considered by architects and designers. It does appear to recognise that standards that might be applied in the real world are either missing or not sufficiently robust, stating that there are no national labelling schemes for building materials or consumer products in England (apart from a scheme for paints). They also note government plans to set up a voluntary labelling scheme in England, as outlined in the government's clean air strategy 2019 while stating a lack of evidence on emissions and claim that:

> Evidence showed that some building materials can emit high levels of pollutants. There was no evidence on building materials and products that emit a low level of VOCs and formaldehyde. **(38)**

The NICE committee agreed that specifying low-emission materials could help protect people's health. But because of the lack of evidence, they could only suggest that professionals consider their use on a case-by-case basis when drawing up specifications.

This is a revealing statement which might raise doubts about the adequacy of much of the whole NICE report. Attending an early consultation "scoping meeting in Manchester on 9 January 2017 and also volunteering when they requested applicants for their committee led to a rather unsatisfactory and far too casual a telephone interview with the future chairman of the committee. Having submitted "evidence" to NICE, responding to their calls through a stakeholder engagement process, and drawing attention to *Building Materials, Health and Indoor Air Quality: Volume 2*, the book was not referenced in the NICE documents.

Individual members of the NICE committee and the officials providing the secretariat were well aware of work on indoor air quality and building materials but seem to have paid it little attention. They failed to address a complaint about the presence in their stakeholder group of a company with commercial vested interests, selling materials which might contribute to poor indoor air emissions. This company went on to figure significantly in the Grenfell Inquiry and remained on the list of NICE stakeholders, as does an organisation representing the wood panel industry. Both these bodies would have had vested interests in limiting guidance from NICE on emissions from building materials such as chemicals in insulation and glues used in composite wood products.

There did not appear to be any involvement from architects or professional architectural bodies such as the RIBA in the NICE stakeholder group.

Despite this, NICE had plenty of opportunity to find out about building materials and products that emit a low level of VOCs and formaldehyde from me. Information was submitted to them about "Natureplus," a highly respected Europe-wide certification system listing hundreds of materials and products governed by threshold limits for harmful substance and emissions into the indoor air, which was listed by the DEFRA. **(39)**

The NICE committee chose to ignore this evidence and indeed appeared not to consider it, even when it was made available to them.

It's hard to determine how much effect, if any, the NICE report has had on local authorities and the construction industry. Despite its flaws, the NICE report raised some red flags about the dangers of mould and an amber flag about hazardous emissions in buildings. Had these warnings been heeded by local authorities and social housing bodies, then the death of Awaab discussed in Chapter 5 might have been avoided. The guide was published in January 2020 with a very simplistic two-page visual summary. Published statements by NICE talk of "limited evidence" about the health impacts of bad indoor air quality. No press reports could be found about the NICE report even in *Inside Housing*, the *Guardian* or other media.

The more detailed evidence published by NICE, which can be found on its website, such as section 3.1 *on Material and Structural interventions,* entirely ignored work on the use of materials that can mitigate mould and asthma and contains a highly unsatisfactory literature review. Building materials are only mentioned twice in 77 pages. Their *Evidence Document 2* on exposure to pollutants and health outcomes almost entirely ignores emissions from materials with the word materials also only appearing twice in 446 pages. While this might reflect the lack of UK epidemiological evidence on the effect of materials emissions on health, there was a strong bias to health effects of cooking and household cleaning products. The evidence document also lists references from scientific work outside the UK but failed to identify such academic work on building materials emissions. **(40)**

A further NICE document has emerged giving guidance for the old "Department of Health" that does at least mention building materials four times in eleven pages stating,

> Indoor air pollutants come from building materials (including fittings and flooring), furnishing, consumer products such as those used for cleaning, candles or diffusers and activities such as cooking and smoking. They also come from biological sources, for example, mould, house dust mites, bacteria, pests or pet dander. **(41)**

As in so many other documents, indoor air emissions from building materials are lumped together with candles, cooking, smoking, etc. This looks like a deliberate attempt to play down the importance of emissions from building materials by confusing them with activities that can be controlled by household occupants.

> NICE (1.6.1) Update existing standards, for example building regulations, or develop new ones for indoor air quality. Base them on current safe limits set for pollutants in residential developments. See, for example, World Health Organization guidelines on selected pollutants (2010) and dampness and mould (2009), and the Public Health England indoor air quality guidelines for selected VOCs (2019). **(42)**

Possible new legislation?

While the wider public and even many local and central government officials may be unaware of the NICE report, there has been a noticeable increase in discussion of air quality in the media and by campaigning groups focussed on traffic pollution. While the majority of this is focussed on ambient air and traffic pollution, the level of interest in indoor air quality has clearly grown since 2017.

There is a Westminster All-Party Parliamentary Group (APPG) on healthy homes and buildings as well as one on air pollution and one on allergies. The APPG on Healthy Homes and Buildings is chaired by Jim Shannon, DUP MP, for Strangford in Northern Ireland. The APPG on Healthy Homes does draw attention to the NICE report and other useful documents on its website. **(42)**

This APPG is funded by Saint Gobain, "Healthy Developments," Velux, BEAMA, Sustainable Energy Association and EON and organised by a lobby group called Devo Connect (now "Inflect"). The APPG has been quite limited in its activities and has failed to address emissions from building materials and the recent serious deaths from mould and damp.

There have been a couple of recent attempts to introduce private members' bills into the Westminster parliament, related to health and air quality such as the "The Healthy Homes Bill," described as "a bill that transforms the regulation of the built environment to ensure that new homes and neighbourhoods support their residents' health and wellbeing."

Initiated by the Town and Country Planning Association (TCPA), and some members of the House of Lords, these proposals initially failed to include any substantive guidance about what anyone reading this book might view as *healthy homes*, as the bill was mainly concerned with outdoor space and general planning and design policies. Following criticism, a small reference to indoor air quality was included in a later version:

> (j) all new homes must not contribute to unsafe or illegal levels of indoor or ambient air pollution and must be built to minimise, and where possible eliminate, the harmful impacts of air pollution on human health and the environment. **(43)**

It seemed unlikely that this bill would survive the Parliamentary process, though it could have been strengthened through collaboration with another bill introduced in the House of Lords by

Baroness Jenny Jones. In September, the TCPA bill had successfully navigated the committee stage in the House of Lords and TCPA argued that the effect of their bill, passed, would be to make healthy homes principles mandatory in housing legislation. While this sounds like a good idea, much more work would need to be done to set out adequate details of what constitutes healthy homes. The lack of detail of what constitutes a healthy home would give house builders a free hand in how to interpret the bill if it ever becomes law. **(44)**

Baroness Jones' Clean Air Human Rights bill does include far more information about indoor air quality with a Schedule 2 on indoor air pollutants and Schedule 3 on pollutants causing primarily environmental harm. It has been referred to as "Ella's law" after Ella Kissi Debra who died of asthma. The coroner ruled that Ella died as a result of asthma aggravated by traffic pollution. **(45), (46), (47)**

Sadly, when this bill was drafted, insufficient technical advice was sought about indoor air quality. While it lists biological indoor air pollutants as dampness and mould it states the rather unrealistic objective that they should be zero! It appears that the bill was drafted with the help of an organisation called Clean Air London, which was established by Simon Birkett. Birkett was a banker who became a campaigner against traffic pollution and is the sole director of Clean Air London (CAL). He is also a director of Client Earth, a firm of lawyers who have been active on the issue of traffic pollution. An article published by CAL in 2012 is the only item on their website addressing IAQ and indicates a fundamental lack of knowledge of the subject, other than reference to the World Health Organisation and a commercial sponsor, an American HEPA filter manufacturer called CAMFIL. **(48), (49), (50)**

Client Earth describes itself as the world's most ambitious environmental organisation with 250 staff in eight offices around the world and high-profile trustees such as the musician, Brian Eno. Client Earth has taken the UK Government to court for failing to meet traffic pollution targets. While it is very active on the subject of air pollution, it has shown no concern for indoor air quality and when contacted denied that it was an issue of concern.

The CAL website lists extensive resources, mainly derived from other organisations, but does not include anything about indoor air quality.

The bill (Ella's Law) lists a set of pollutant-specific guidelines for chemical pollution (see footnote) which is a step in the right direction, taking up the NICE advice for greater regulation, but it isn't clear as to why these particular chemicals were chosen, though Section 3.2.c seems to suggest that they were derived from the World Health Organisation.

The bill proposes that the Secretary of State should establish regulations for the sampling, measurement and reporting of air pollutants (but not indoor air) and developers should assess and report concentrations of indoor air pollutants without providing any guidance on how this might take place.

Given the lack of joined up thinking in the UK Parliament, it is not surprising that the Government introduced yet another piece of legislation in 2023, unrelated to the two private members bills, known as the Housing Safety, Social Housing Regulation bill. This was amended following the death of Awaab in Rochdale and became referred to as Awaab's law:

> On 9 February 2023, Michael Gove announced the Government had tabled amendments to the Social Housing (Regulation) Bill to introduce 'Awaab's Law', which would require landlords to investigate and fix damp and mould in their properties within specified timeframes. **(51)**

This bill could have been an opportunity to respond to the NICE recommendations, linking up with the TCPA and Jones bills by introducing limits of hazardous emissions from building materials, but this does not appear in the proposed new regulations, which instead were largely

Table 4.3 Extract of limits in a schedule from the Clean Air Human Rights Bill

SCHEDULE 2 Section 1

INDOOR AIR POLLUTANTS

1 Biological indoor air pollutants (dampness and mould) Zero

2 Pollutant-specific guidelines (chemical pollution)

Pollutant Concentration Averaging period

Dampness Zero n/a

Mould Zero n/a

Pollutant Unit Averaging period

Benzene (C6H6) 3.5 µg/m3 1 year

1, 3 Butadiene 2.25 µg/m3 1 year

Carbon monoxide (CO) 4 mg/m3 24 hours 10 mg/m3 8 hours, 35 mg/m3 1 hour, 100 mg/m3 15 minutes

Formaldehyde (HCHO) 8.6 µg/m3 1 year

Hydrogen sulphide (H2S) 7 µg/m3 30 minutes 0.15 µg/m3 24 hours

Nitrogen dioxide (NO2) 200 µg/m3 1 hour (no exceedances) 25 µg/m3 24 hours 10 µg/m3 1 year

Naphthalene 0.01 mg/m3 1 year

Polycyclic aromatic hydrocarbons (PAHs)expressed as concentration of benzo(a)pyrene Zero n/a

Radon 100 becquerels/m3 3 months

Tetrachloroethylene 0.25 mg/m3 1 year

Trichloroethylene Zero n/a

Schedule 2 includes limits to indoor air pollutants, but these are expressed purely as particulates

concerned with methods and equipment for surveying housing defects, and how landlords might behave towards their tenants.

While this Government-backed bill was going through Parliament an advisory expert group was hastily convened to produce guidance on damp and mould in buildings with plans to issue a guidance document in July 20123. This is discussed in more detail in Chapter 5.

Distracting attention from IAQ using cooking and wood burning stoves

One of the more curious attempts to deflect attention from emissions from building materials has been recent statements about emissions from cooking, claimed as one of the most serious sources of bad indoor air quality. This has been encouraged by the DEFRA AQEG report and to a lesser extent the NICE report. Having largely ignored emissions from building materials, the DEFRA report mentions emissions from cooking 61 times, and also contains numerous references to wood burning, stoves and fires.

It is difficult to discern how this pre-occupation with cooking as an indoor air problem came to prominence in the last few years but recent academic statements about the problem were rapidly picked up by newspapers. A number of companies promoting air purifiers and ventilation systems spotted this as a marketing opportunity and cooking with woks, in particular, became a target, following a study from Singapore:

> Research into the impact of wok cooking found that it leads to significantly increased levels of the toxicants acrolein and crotonaldehyde, substances which can attack a person's DNA. The complicating issue for many of these women was that because they were not thought to be in a risk group for lung cancer, the diagnosis came late. It's a shocking discovery, one of several that have brought an increasing focus on the quality of the air we breathe inside our homes. **(52)**

Evidence about the dangers of cooking draws on problems from particulates and NO2 emissions, but particulates can come from a number of indoor sources, not necessarily cooking. There is no doubt that health problems can result from cooking on open fires in poorly ventilated huts and shacks in developing countries with the World Health Organisation drawing attention to this problem, but data on health problems from Africa have been misused to attack gas cooking in the UK.

- Around 2.4 billion people worldwide (around a third of the global population) cook using open fires or inefficient stoves fuelled by kerosene, biomass (wood, animal dung and crop waste) and coal, which generates harmful household air pollution.
- Household air pollution was responsible for an estimated 3.2 million deaths per year in 2020, including over 237,000 deaths of children under the age of 5.
- Worldwide, around 2.4 billion people still cook using solid fuels (such as wood, crop waste, charcoal, coal and dung) and kerosene in open fires and inefficient stoves (1). Most of these people are poor and live in low- and middle-income countries. **(53)**

Gas cooking and other fuel sources such as log burning stoves can be a problem in terms of carbon monoxide (CO) poisoning, but it was interesting to find how the anti-gas cooking literature rarely refers to CO. **(54)**

There have been nearly 700 deaths from CO poisoning in the UK in the last five years, but numbers are dropping as awareness has improved, with more CO detectors installed. Lower levels of CO poisoning can also have serious health effects.

[Footnote. The author has been chair of the All Party Parliamentary Carbon Monoxide stakeholder group since 2021.]

When giving evidence to the Westminster Parliament Environmental Audit Committee on July 5, 2023, Larissa Lockwood from Global Action Programme (GAP) and Professor Nicola Carslaw, from University of York, made strong claims about the dangers of gas cooking and argued that all cooking should be done on electric induction hobs. Lockwood cited "evidence" from "CLASP" that indoor air pollution from gas cooking costs the UK around £1.4 billion a year in healthcare costs.

There are some serious flaws to these arguments. First of all, a good cooker extract fan should be used when cooking and the vast majority of emissions from that process will be extracted limiting effect on the indoor air quality. However, not all houses have such extract fans (Chapter 6). Professor Carslaw stated, at the Environmental Audit committee, that frying meat on a gas hob created the highest levels of toxic emissions possible in a house (300 micrograms per metre cubed, 60 times the WHO safe limit) and that electric hobs should be used instead. However, frying foods on electric stoves can actually produce higher particulate levels than gas stoves, according to research in California. **(55)**

Opposition from a number of scientists and academics to cooking on gas seems commonplace but possibly driven by decarbonisation policies to replace gas usage with electricity.

Hazardous emissions from the PFAS in "non-stick" cooking pans may be far more dangerous than gas, as can cleaning materials which are often used on hot surfaces such as induction hobs. Nitrogen dioxide emissions from gas cooking is a health hazard and can cause respiratory problems and so should not be ignored, but the anti-gas cooking lobby mainly focusses on particulates, paying little attention to other chemical emissions. **(56)**

Lockwood also spoke about emissions from burnt toast and stated that levels of air pollution in homes were about 500 times levels outside when people were cooking. She said that this was based on work with National Air Quality Testing Services (NAQTS). **(57)**

NAQTS has an official sounding name but is in fact a private company run by three members of the Booker family, with an office in Ann Arbor Michigan USA and links with Lancaster University Environment Centre. It was not possible to find any company registration details for NAQTS in the UK. NAQTS claims to provide indoor air quality monitoring services but nothing could be found on their website to back up Lockwood's claims.

Lockwood's organisation, Global Action Plan, cites the work of an organisation called CLASP which campaigns to replace gas with electricity. However, when searching the CLASP website for research on gas cooking, CLASP, which claims that indoor air pollution from gas cooking costs the UK around £1.4 billion a year in healthcare costs, it turns out that they are quoting the work of GAP which in turn claims to be based on the work of CLASP! **(58), (59), (60)**

CLASP is a global organisation with funding from a wide range of corporate and charitable sources including Modern Energy Cooking Services at Loughborough University **(61),** which provides help to people in 16 countries in Africa and Asia.

Clasp has bases in USA, New Delhi, Jakarta and Nairobi and state that they are promoting global electrification. Their 2022 annual report gives an address in Washington DC and shows funding from the Rockefeller Foundation, The World Bank and the United Nations among many other big foundations. This organisation may be doing good work to assist people in poorer countries with "clean" energy, as they work with the banking sector to fund renewable energy. However, their claim to have accumulated 400 research publications including work by Professor Frank Kelly at Imperial College, who is a trustee of GAP, does not provide a list of these on their website.

There is a major campaign throughout Europe against the installation of gas cookers **(62)** with the work of Steffan Loft frequently cited. Loft is a joint author within a paper by Lim et al on indoor air quality. **(63)** However, the research by Lim and Loft et al suggests that the levels of exposure to ambient air pollution and secondhand smoke are much more harmful than are the levels of exposure to indoor combustion sources from candles and wood stoves in a high-income setting. Cooking from gas barely merits a mention!

The CLASP report on electrifying cookers cites EU data on health problems from indoor air pollution, but fails to separate out how much of that is due to gas cooking. Indeed, their claims are not based on actual emissions data at all, but "modelling and simulation studies." They say that cooking with gas releases the following pollutants, nitrogen dioxide, carbon monoxide, nitrogen monoxide, methane, ultra-fine particles and PM 2.5. They estimate the number of children with asthma due to cooking with gas (with Italy having the highest levels; UK is not included) costs the EU €3.5 billion in health care but these seem to be based on genera asthma rates that may have been caused by many other factors such as mould and damp and emissions from building materials and may not be directly related to cooking.

While cooking with cleaner electricity may be safer and preferable (though not cheaper), the evidence of health harm from gas cooking is flimsy to say the least, and the campaigning around this issue looks more like propaganda from the electricity industry. However, most of the campaigning by the electricity industry on gas has focussed on heating with gas and they have been successful in persuading the UK Government to ban all gas boilers in new build homes from 2025, though the current Conservative government looks likely to delay this policy. New build homes may be less likely to be connected to a gas network and homes will have to use electricity for cooking. When coupled with scare stories about wood burning stoves (discussed below) in articles and studies, which barely mention other sources of indoor air pollution such as chemicals, further research is required before firm conclusions can be drawn. It is claimed that the Gas industry knew about health harm from cooking with gas 50 years ago and tried to cover up the evidence and a draft report from AGA published in 1972 was followed up with by the National Industrial Pollution Control council in the USA. AGA do supply stoves fuelled by gas, wood and solid fuel. **(64) (65)**

As pointed out by Nicola Davis, the Guardian Science Correspondent on May 18 2023, prehistoric humans may have been "toasting snacks" on open fires as long ago as 250,000 years ago and maybe even earlier in Africa and parts of Europe. It is possible that cooking on wood stoves and gas may have been causing health problems for generations, but scientific analysis seems to have been overtaken by electricity campaigning. **(66)**

It is necessary to ask the question why health scare stories have emerged in relation to cooking that were not previously regarded as a problem for thousands of years. DEFRA AQEG also cited cooking as a major course of indoor air pollution. The AQEG report does admit that claims of adverse health effects from cooking are largely based on problems in the developing world, but undaunted they go on to devote considerable space to cooking such as "smoke and sizzling."

Wood burning stoves

DEFRA AQEG immediately follow their warnings about cooking by discussing emissions from solid fuels. As with cooking they cite research on adverse health effects coming from work in developing countries, with their main reference for emissions from residential heating stoves from a PhD student at Sheffield University. This study by Chakroborty and Heyden provides negligible data on emissions from stoves and was focussed instead on what they called "participatory sensing." **(67)**

While there are some indoor health risks from cooking a great deal more attention has focussed on pollution from wood burning stoves. Professor Sir Chris Whitty in his evidence to the environmental audit committee in Parliament on July 5 states a figure of 17% of all particulate emissions resulting from wood burning stoves. Rather, as with cooking, newspapers have promoted many scare stories about pollution from stoves, and initially ambient emissions from wood burning were said to be responsible for 38%, a figure revised down when data was requested from various official bodies for this book. Figures about emissions are based on estimates from industry sales figures on how much solid fuel had been used including coal as well as wood (and not from atmospheric data). When sampling external air, it requires a great deal of sophisticated equipment to isolate different sources of particulates and chemicals, thus separating out emissions from wood burning stoves is not easy. The data is confused with emissions from garden fires and barbecues, for instance. **(68)**

It remains difficult to find accurate figures of emissions from wood burning stoves despite claims that it is a massive problem. DEFRA's figures are derived from a *market research* company organisation called Kantar, which is a global data and marketing consulting company, providing information on a wide range of topics from LGBTQ in advertising, brand tracking and audience measuring. DEFRA commissioned Kantar to estimate how much solid wood fuel was being burned, but nothing on emissions from stoves could be found on the Kantar website. The Kantar report states that only 8% of UK households burn solid fuels in the winter and only 1% in the summer. The Kantar work also does not address the issue of emissions and their data does seem to suggest that claims of massive emissions from wood burning have been significantly overexaggerated. **(69)**

In his talk at a Clean Air conference in Cardiff, Dr. Gary Fuller, a respected scientist from Imperial College, quotes the DEFRA Kantar research to claim that 27% of particulates from solid fuel is nearly twice the emissions from transport exhausts, which is a strange statistic as he does not explain what happens to the particulates from the other 73%. **(70)**

Fuller, a "Clean Air Champion," and a member of the DEFRA AQEG, has been a leading campaigner against wood burning stoves and has regularly written in the *Guardian* newspaper, claiming that wood burners emit more particle pollution than traffic and that a rise in UK wood burners is likely creating "pollution hotspots" in affluent areas. He is highly critical of middle class people for installing stoves for what he calls aesthetic and lifestyle reasons. **(71), (72)**

As with cooking, this raises some curious contradictory issues, because affluent homeowners installing wood burning stoves are much more likely to have them installed by reputable good quality certified stove companies. If they are there for purely aesthetic reasons then it's likely they are not using the stoves very much anyway, but to suggest that "eco: wood stoves emit 750 times more pollution than an HGV, as does Damien Carrington in the *Guardian*, is patently absurd. He claims that unnamed "experts" say that stove regulations are shockingly weak and that wood burners triple the levels of pollution in the home with very little evidence. **(73)**

Fuller also claims that many people are burning waste wood and that this is causing a release of arsenic into urban air. He cites arsenic measurements in a graph from the "London Supersite" by Tremper and Trechera, colleagues at Imperial, though the published source of this data is not made clear by Fuller. **(74)** Fuller's implication that burning waste wood is leading to arsenic emissions is curious as Copper Chrome Arsenic (CCA) treatment of timber was banned in 2006, and while there might be some old scrap treated wood that is still being collected today, any treatment emissions are likely to be from other chemicals such as phosphates, not arsenic. Accurate data on emission levels from burning chemically treated wood should be considered as a serious hazard, but the real danger to householders is from the emission levels from treated timber inside homes, an issue not referred to by Fuller. **(75)**

Solid fuel is still burned in open fires and some people may be using cheap wood burning stoves that do not meet current regulations, but if good-quality, air-tight, HETAS-approved **(76)** certified stoves burn good quality dry wood, then there may be some emissions from chimneys and flues, but particulate emissions inside houses should be limited. With the rise in gas and electricity costs, many more people have resorted to burning solid fuel and possibly even rubbish in an effort to save money, and this may have accounted for these assumptions about an increase in emissions. It is hard to avoid the conclusion that the hysteria about wood burning stoves is yet another intent to distract attention from building materials emissions.

Particulates

Levels of particulates in external air can vary considerably and is much higher in certain areas of congested roads and places like Birmingham New Street station, (from diesel train engines left idling) for instance, but current monitoring needs to be extremely sophisticated to separate out particulates from traffic, industry and wood burning. The published work by Chakraborty and Heydon contains negligible data on emissions from stoves and instead relies on statements by DEFRA who have then cited Chakraborty. He now works for a private air quality assessment company called AirRated Ltd and Heydon is a criminologist. Their published work was based on low-cost sensors.

> Accounting for 17% of PM2.5 emissions nationally, PM from domestic burning increased by 35% between 2010 and 2020 (Department for Environment, Food and Rural Affairs 2022). After four weeks of participatory sensing, where laypersons used sensors to identify indoor air quality during stove use, the results show how monitoring technology pulls wider preconceptions into the data interpretation process. When faced with numerical data perceived as ambiguous, users draw on preconceptions that frame stoves in a positive light. **(77)**

However, at the Cardiff conference on June 28, 2023 at the Principality Stadium Cardiff (unpublished) Chakraborty showed some limited data on emissions and suggested that wood burning stoves are bad for indoor air quality creating high levels of particulates indoors. This is derived from data from sensors placed in the vicinity of 20 stoves over 4 weeks with peak concentrations of 27.34 ug/m3, though not always this high. He claimed that PM 2.5 and PM 1 levels soar by around 200% when stoves are being refuelled, and if this data is correct, based on this very small study, then stoves can be seen as responsible for worsening indoor air quality. But the hyperbolae used by these academics such as "Burner alert," "Call to Action," "Particulate flooding" and "PM hurricanes" must raise questions about whether this is based on objective scientific analysis or personal prejudice.

Another speaker about wood burning emissions at the Cardiff conference, Wenger, is a joint author with Byrne, of a major study of particulate emissions in which PM 2.5 was a monitored by "low cost" sensors across the city of Cork in Southern Ireland, and mentions air quality being impacted by solid fuel burning in many countries. The paper refers to emissions from wood burning or stoves using chemical markers with discussion of the fingerprints of fuel emissions and seems much more detailed and comprehensive than other studies already referred to. One of the Cork findings was interesting, in that higher particulate levels were found in the countryside rather than in the city centre, due to higher instances of residential fuel burning during months like February and high levels were found at 9 of their 24 monitoring stations. In Ireland, according to the Irish Environmental Protection Agency, 38% of homes use solid fuel as supplementary heating, mostly coal and peat. **(78)**

The subject of stoves and wood burning is complicated and worthy of a much more detailed analysis than is possible in this book. The main point to be made here is that evidence of emissions from wood burning is still patchy and may be being used as a "smokescreen" to draw attention away from chemical emissions from building materials.

> Some people hate the scent of woodsmoke. They think it's a 'mucky' smell but I love it…… .To me it's a safe, rich homely smell that evokes the scent of childhood and all the fires, that flickered and glowed throughout my early years. **(79)**

> Who has smelled the woodsmoke at twilight, who has seen the campfire burning, who is quick to read the noises of the night?
>
> (Rudyard Kipling)

Public Health England indoor air quality guidelines

Public Health England is now part of the UK Health Security Agency and Office for Health Improvement and Disparities. It issued indoor air quality guidelines in 2019 based on an academic paper by Shrubsole and Dimitroulopoulou, et al. This document takes a different approach from the other policies adopted by DEFRA and NICE and is referred to in consultation documents about changes to the UK building regulations, in which it is suggested that these PHE guidelines would be used in the regulations. The PHE document points out that there are no indoor air quality guidelines for individual VOCs. Part F of the English regulations, which provides ventilation standards (2010), sets maximum concentrations of TVOCs and quite rightly the PHE document states that,

> TVOCs reveal little regarding the nature of the individual compounds, their concentrations and their possible toxicity to humans. Although there is a plethora of pollutants found indoors,

> including gaseous pollutants (inorganic chemicals, radon, volatile organic compounds, VOCs), biological pollutants (allergens, viruses and bacteria, mould) and particulate matter, the current work focusses on indoor generated VOCs and the guideline values that would aim at their control in the indoor environment. VOCs are widely used in construction and building products. **(80)**

The PHE team proposes air quality guidelines for a limited selection of eleven VOCs emitted from construction products and building materials, based on a literature review of 30 references, some of which were written by the authors of the report. These are Acetaldehyde, A-Pinene, Benzene, D-Limonene, Formaldehyde, Napthalene, Styrene, Tetrachloroethylene, Toluene, Trichloroethyeleme and Xylenes. They include these in a table setting out short- and long-term limit values in ug/m3 derived from USA, Canada, WHO and other sources. They also provide a good explanation of health impacts. The suggestion that this list should set VOC limits in the UK building regulations seems like a step in the right direction as other Government agencies have largely ignored the issue of building materials emissions. However, there is a danger that this list will be seen as the last word in terms of building regulations and has already been cited in some draft changes to the building regulations. The PHE report fails to address the issue of how these limits would be applied and how buildings would be tested to indicate compliance. Also, there is no explanation as to why these particular chemicals were chosen, and other dangerous chemical emissions from potentially more dangerous substances such as iso-cyanates, flame retardants and many more listed were omitted.

Shrubsole and Dimitroulopoulou et al have published a more extensive paper in which they say that "individual VOC guidelines are the most appropriate way forward" without making clear forward to what. They also conclude that TVOC can be used as an indicator for indoor air quality, which seems to contradict their previous statement. Their selection of the 11 substances is justified on the basis of prioritising these for monitoring purposes. **(81)**

The discussion within the world of building regulations is largely focussed on ventilation though a recently issued consultation document in Northern Ireland also refers to work in Scotland where it has been suggested that monitoring CO2 levels in houses should be standard. **(82)**

The Scottish building regulation authority's requirement of CO2 monitoring in the principle bedroom of each house, is on the basis that this can act as a proxy for indoor air quality in dwellings. The CO2 detection is not installed as an alarm but in order that occupants can check whether they have enough ventilation. This concept of CO2 as a proxy for VOCs is discussed in UK indoor air quality literature but has not been subject to proper review. IQ Air confirm that there is <u>no</u> direct correlation between indoor CO2 and pollutants such as VOC. CO2 levels shown on sensors could give a false sense of security to occupants that VOC levels are ok when they may well be at dangerous levels. This seems to be yet another attempt to avoid dealing with hazardous chemical emissions from building materials as they will not be picked up by CO monitors.

> There is no direct correlation between indoor CO_2 and other common indoor air pollutants, such as particulate matter (PM) **or VOCs**. In some cases, indoor CO_2 may exhibit behaviour opposite to that of other indoor air pollutants. For example, opening a window on a polluted day may reduce indoor CO_2 but increase <u>PM10</u>, <u>PM2.5</u>, and other <u>outdoor air pollutants</u> that penetrate the indoor space. However, conditions that lead to high levels of CO_2 can also increase indoor concentrations of PM or VOCs. In a poorly ventilated or unfiltered space,

both CO_2 and other indoor air pollutants can build up to dangerous levels and result in a wide variety of health effects. **(83)**

UK building regulations and indoor air quality

There is some recognition that increased pressure to improve thermal performance and air tightness in dwellings can have a negative effect on indoor air quality, but organisations like the Passivhaus Trust (PHT) are lobbying hard for even more extreme levels of air tightness and a reliance on mechanical ventilation to maintain good air quality. **(84)**

The PHT approach rarely considers the issue of emissions from the plastic and hazardous building materials used to achieve passivhaus standards. A number of lobby organisations, who advocate changes to building regulations, advocate the adoption of passivhaus standards, such as the London Energy Transformation Initiative (LETI). The much quoted 158-page LETI Climate Emergency Design Guide, contributed to by over 100 people, only mentions indoor air quality three times, mentioning emissions from OSB but advocating mechanical ventilation (MVHR) as the only solution to good air quality. **(85)**

Standard assessment procedure

The UK Government department BEIS (the Department for Business, Energy and Industrial. Strategy, now split (from 2023) into the Department for Energy Security and Net Zero, Department for Science Innovation and Technology and the Department for Business and Trade, commissioned an organisation called ETUDE to review what is known as the Standard Assessment Procedure (SAP), a cornerstone of energy efficiency measures. Anyone submitting a building design for approval under the building regulations has to submit a SAP. The SAP is primarily concerned with energy efficiency but its many and complex requirements have a significant effect on the selection of materials and other key design features that can affect indoor air quality. Awbi has suggested oncreased SAP ratings have been linked to an increase in indoor air pollution and health problems. **(86)**

The ETUDE study, which involved extensive consultation, and an input from many UK and international experts, came up with 25 recommendations which includes a suggestion that SAP should better model the energy efficiency of ventilations systems (note this is energy efficiency, not effectiveness!). They say that SAP should reflect all energy uses including white goods and cooking and take account of the location of houses. An opportunity for SAP to take account of indoor air quality was entirely missed. Etude, with its close links to Passivhaus, reinforced the pre-occupation with extreme energy efficiency without considering health impacts. The study did include a working group on Ventilation and Indoor Air Quality and the Appendix F in their report does mention “evaluate indoor air quality” as part of a discussion of Health and Wellbeing, but without any explanation of the significance of IAQ and no proposals for the revision of SAP relate to indoor air quality. **(87)**

One of the organisations playing a key role in this work was the Clarion Housing Association, which has attracted severe maladministration findings from the UK Housing Ombudsman. Another key partner was Levitt Bernstein Architects who have worked on many Clarion projects. Levitt Bernstein are closely linked to Etude. **(88)**, **(89)** Etude, the lead organisation in this study, is a relatively new firm of engineers established in 2012, who share an office space with the Passivhaus Trust in Islington in London and prominently display their passivhaus allegiance on their website. **(90)**

Currently it looks unlikely that there will be any limits on indoor emissions in the building regulations. A study carried out by consultants AECOM for the Ministry of Housing Communities and Local Government (also no longer in existence) called Ventilation and Indoor Air Quality in New Homes provided a limited study of emissions in terms of NO2, *body odour* and TVOCs. Their research revealed that 6 out of the 10 bedrooms they studied had TVOC concentrations higher than the standard of 300 µg/m3. However, the report claims that current ventilation regulations are appropriate for controlling NO2, carbon monoxide and formaldehyde. This seems to be a highly dangerous conclusion to reach on the basis of a tiny study of 10 houses, however to their credit, AECOM do state that it would be useful to consider the benefit of source control, as lower pollutant emissions would require less ventilation and reduce the building energy consumption. They also admit that formaldehyde and TVOC concentrations are 20 times higher inside than out and they criticised building regulations guidance, stating

> "Source control is not considered within the main guidance of the Approved Document owing to limited knowledge about the emission of pollutants from construction products used in buildings and the lack of suitable labelling schemes for England and Wales" **(91)**

Finding satisfactory guidance for indoor air quality?

As should be clear from this book, establishing safe levels for chemical emissions in buildings is a massive challenge. The world of indoor air quality is full of conflicting and confusing science, standards and organisations. There is little evidence of the many experts and organisations attempting to get together and agree on a common approach and bodies such as the United Nations Environment Programme (UNEP) or the World Health Organisation (WHO) do not seem to have attempted this. UNEP is quite rightly concerned about indoor cooking in very poor countries but doesn't provide any guidance on chemical emissions in modern buildings. **(92)**

In the UK, the recently established Office of Environmental Protection (OEP) **(93)** has so far failed to properly tackle the issue of indoor air quality. In its recently published "Stocktake Report," produced for it by an organisation called Ricardo, it says that understanding of indoor air quality *"is in its infancy,"* presumably as they haven't bothered to read the many thousands of books and papers on the subject! The OEP says that the Ricardo report does not reflect its views and the infancy comment was a "summary of what other scrutiny bodies and organisations were found to have said during their research." **(94)**

> 4.3.5 Evidence of the impact of indoor air quality
> It is generally acknowledged that our understanding of indoor air quality is in its infancy and further research into this area is commencing. For decades Government action has been on ambient outdoor pollution levels, driven by regulations. Occupational exposure to high pollutant levels has been controlled and enforced under relevant health and safety legislation which has left indoor air quality in homes and other public spaces largely unregulated.
>
> Building design is also important, and while ventilation is a crucial consideration to meet energy efficiency targets, the impact on air quality and respiratory infections should also be considered as a potential trade off.
>
> More evidence is being published on volatile organic compounds emissions from products regularly used within residential homes. **(95)**

The UK Government subcontracts a great deal of work on air quality and other environmental issues to Ricardo, a private global corporation, which works on aerospace and defence, energy utility and transport issues. Initially a UK company, with a base in Shoreham by Sea it is worth over £400 million and leads a consortium delivering a £5 million research programme on climate change. Ricardo is partnering with University College London (UCL); Tyndall Centre for Climate Change Research and institutes supported by the Natural Environment Research Council (NERC), including the British Antarctic Survey (BAS), British Geological Survey (BGS), National Centre for Atmospheric Science (NCAS), National Centre for Earth Observation (NCEO), National Oceanography Centre (NOC), Plymouth Marine Laboratory (PML) and the UK Centre for Ecology & Hydrology (UKCEH).

Ricardo seems to dominate the field in terms of data on air pollution in the UK as it is frequently referenced in policy documents. However, as the OEP report reveals, Ricardo does not deal with indoor air pollution. **(96)**

International Society of Indoor Air Quality

International academic bodies such as ISIAQ provide a useful forum for academic research. The ISIAQ provides a comprehensive list of IEQ guidelines from different countries around the world in a database. **(97)**

There is also a very useful map where it's possible to click on countries to get a summary of how much standards can vary, with TVOC levels limited to 200 in Spain, but 500 in South Korea and in the WELL Standards. Many countries do not even set TVOC guidelines. Germany sets a much more detailed set of guidelines than most other countries.

The Institute of Air Quality Management

The IAQM have published a very useful, though rather general, guidance document on indoor air quality though it is very limited in its discussion of emissions from building materials. It includes "Table of Indoor Air Quality Assessment Levels" which is so general that it is not very helpful. However, the document does cite the work of this author as "thought provoking." **(98), (99), (100)**

The IAQM is part of the Institution of Environmental Sciences which is a registered charity and publishes *Environmental Scientist* which occasionally focusses on indoor air quality.

The IAQM refers to indoor air quality in the BRE Home Quality Mark, which contains little about indoor air emission, though it is linked with BREEAM which does contain standards for indoor air. It avoids discussing standards or limits to VOC emissions as does the BRE.

> Inside the home: Sources of pollution inside the home are often the choice of the occupant (cleaning products, furniture, cosmetics etc). However, the construction materials, including paints and varnishes can have a significant impact of levels of Volatile Organic Compounds (VOC) and airborne Formaldehyde. HQM awards for construction materials that have limited impact upon indoor pollution. **(101)**

World Health Organisation

The UK Government, the European Union, the US Environmental Protection Agency and ASHRAE, WELL and the WHO all have different approaches but tend to refer to the WHO.

WHO has published a number of important publications about indoor air quality, but contrary to general belief has not set clear standards for indoor emissions that can be adopted by various countries such as the UK. The WHO Global Air Quality Guidelines published in 2021 only addresses particulates: Ozone, NO2, SO2 and carbon monoxide. The word indoor does not appear once in its 300 pages. On the other hand, the excellent WHO report, "Measures to reduce risks for children's health from combined exposure to multiple chemicals in indoor air in public settings for children with a focus on schools, kindergartens and day-care centres," refers to indoor air over 100 times and raises considerable concern about exposure to multiple chemicals in indoor air and their effects on child health.

> 2.2.1. Chemicals originating mainly from indoor sources
> A wide range of airborne chemicals can be emitted from multiple sources inside educational buildings due to construction, renovation, operation and maintenance, as well as from materials used for educational purposes and during certain activities.
>
> 2.2.1.1. Continuous emissions
> The common groups of chemicals in indoor air in educational settings for children are aldehydes, VOCs (aromatic hydrocarbons, esters, terpenes and chlorinated hydrocarbons) and semi-VOCs (brominated flame retardants, PAHs, perfluorinated compounds, phthalates). Aldehydes and VOCs are of high concern due to their ubiquitous presence.
>
> The most investigated chemical pollutants that are widely present in school environments and significantly affect children's health are formaldehyde, benzene, ethylbenzene, toluene, xylenes, naphthalene, styrene, limonene and α-pinene. A re-analysis of the data of the Europe-wide SINPHONIE project showed that 29%, 19% and 11% of schoolchildren were co-exposed to elevated concentrations (> median value) to 2, 3 and 4 VOCs known to be harmful for health, respectively. Most of these compounds are usually found at higher concentrations indoors than outdoors.
>
> 2.2.2.1. Source control
> The most effective mitigation measures are those that focus on the elimination of indoor sources of harmful pollutants... The use of low-emitting products will reduce continuous emissions into indoor air. **(102)**

A related WHO publication produced in Europe, "Chemical pollution of indoor air and its risk for children's health" provides an educational course on different aspects of chemical pollution of indoor air and its risk for children's health, including principles of risk assessment. While this recognises the significance of building materials as pollutant sources and is very well referenced and lists many chemical pollutants, it could have gone into more detail about the nature of building materials which are sources of pollution, such as flame retardants, which are not mentioned in the section on endocrine disruption. It provides a great deal of detail on health effects on children but fails to address the problem of mould and dampness. **(103)**

WHO produced a guide to selected indoor air pollutants in 2010, but this only addresses nine pollutants: Benzene, Carbon Monoxide, Formaldehyde, Napthalene, Nitrogen Dioxide Polycyclic aromatic hydrocarbons, Radon, Trichlroethylene and Tetrachhloroethylene. **(104)**

> The evidence review supporting the guidelines for each of the selected pollutants includes an evaluation of indoor sources, current indoor concentrations and their relationship with outdoor levels, as well as a summary of the evidence on the kinetics and metabolism and health effects. Based on the accumulated evidence, the experts formulated health risk evaluations and agreed on the guidelines for each of the pollutants as summarized below. **(105)**

Table 4.4 Summary of indoor air quality guidelines for selected pollutants derived from WHO

Benzene	Acute myeloid leukaemia (sufficient evidence on causality) • Genotoxicity
Carbon monoxide	Acute exposure-related reduction of exercise tolerance and increase in symptoms of ischaemic heart disease (e.g., ST-segment changes)
Formaldehyde	Sensory Irritation? No mention of it being a suspected carcinogen!
Napthalene	Respiratory tract lesions leading to inflammation and malignancy in animal studies
Nitrogen dioxide	Respiratory symptoms, bronchoconstriction, increased bronchial reactivity, airway inflammation and decreases in immune defence, leading to increased susceptibility to respiratory infection
Polycyclic aromatic hydrocarbons	Lung cancer
Radon	Lung cancer Suggestive evidence of an association with other cancers, in particular leukaemia and cancers of the extrathoracic airways
Trichloroethylene	Carcinogenicity (liver, kidney, bile duct and non-Hodgkin's lymphoma), with the assumption of genotoxicity
Tetrachloroethylene	Effects in the kidney indicative of early renal disease and impaired performance

A WHO working group met in Bonn in 2006 to try and develop specific indoor air guidelines, but these do not appear to have emerged; however, it did lead to useful guidance on mould and damp. **(106)**

WHO published new air quality guidelines on September 21 2021, greeted with great acclaim by some organisations, but these guidelines only refer to particulates and say nothing at all about indoor air pollution. The word indoor does not appear once in 300 pages. **(107)**

> Some academics have suggested that the ambient air limit values set by WHO "can mostly be translated into indoor limit values." The built environment needs to harmonize energy efficiency, thermal comfort and air quality standards and guidance. In this review, we discuss the next steps for improving international, regional and national standards and guidance, leading to better and more complete indoor air quality regulations. **(108)**

BSI, ISO and CEN

The British Standards Institute and the International Standards Organisation have considered indoor air guidance but have not produced a clear set of standards. BS 40102-1 gives "*recommendations*" for measuring and reporting on the indoor environmental quality (IEQ) of non-domestic buildings and claims to be a rating system.

The BSI standards are usually closely linked to standards produced by CEN and ISO such as CEN 16798-2 and ISO 7730. CEN is the European Committee for Standardisation. CEN acts as host for hundreds of committees and technical working groups and issues reports and standards. **(109)**

Table 4.5 BS 40102-1 includes (among other things)

5.2 Particulate matter
5.3 Carbon monoxide
5.4 Total volatile organic compounds
5.5 Nitrogen dioxide
5.6 Ozone
5.7 Carbon dioxide
7 Thermal comfort
7.2 Naturally ventilated buildings
7.3 Additional considerations for buildings with fan powered ventilation or air-conditioning
8 Acoustic and soundscape quality

CEN covers a huge range of topics but surprisingly has little on indoor air quality and emissions from materials. When it does, it tends to fall under sustainable construction and energy performance of buildings. CEN works closely with global construction research body, International Council for Research and Innovation in Building and Construction (Conseil International du Bâtiment CIB. **(110)**

The CEN report "Energy performance of buildings — Ventilation for buildings – Part 2" is an example of a technical report which establishes design criteria for long-term evaluation of indoor environments and parameters to be used for monitoring indoor environmental conditions, though it has not been possible to find any case studies demonstrating whether this has been used at all. The focus for these standards and the work of various ISO groups seems largely centred on energy and ventilation. **(111)**

The only discussion of 16798-2 was found in a paper by Oleson which confirms that good indoor environmental quality is seen as an issue of heating, cooling and ventilation by and large. **(112), (113)**

There are a number of other ISO standards, ISO TC163 and ISO TC205 ISO 17772-1, and TR 17772-2 ISO 16814:2008 ISO standard ISO 16814:2008 includes the following:

> This document does not prescribe a specific method but rather refers to existing methods in published standards and guidance, ... the aim of the methods is to control indoor air pollutants to concentration levels below which, under the prevailing hygro-thermal conditions, the pollutants do not have the potential to
>
> - cause a significant risk of adverse health effects,
> - adversely affect the comfort of the majority of occupants.
>
> The pollutants considered include human bioeffluents, which have often been the principal consideration for IAQ and ventilation, but also all groups and sources of pollutants that can reasonably be anticipated to occur in the building being designed. The pollutants to be considered can, depending on the sources present, include
>
> - volatile organic compounds (VOCs) and other organics, such as formaldehyde,
> - environmental tobacco smoke (ETS),
> - radon,
> - other inorganic gases, such as ozone, carbon monoxide and oxides of nitrogen,

- viable particles, including viruses, bacteria and fungal spores,
- non-viable biological pollutants, such as particles of mites or fungi and their metabolic products,
- non-viable particles, such as dusts and fibres.

This international standard is largely concerned with airflow/ventilation rates and contains a dilution index. Another ISO standard on emissions rates ISO 4225:1994 has been replaced by 4225:2020 which includes an indoor air quality standard. **(114)**

ETSI Working Groups

It has published 943 European standards in 2022, 146 technical specifications and there are 16,436 standards in total with many more to come.

Despite this plethora of work, there are no standards on indoor air quality though there is a relatively dormant group CEN TC351 which has produced CEN/TS 17331:2019, which specifies existing (not new) methods for the determination of the content of specific organic substances including BTEX, biocides, dioxins, furans and dioxin-like PCBs, mineral oil, nonylphenols, PAH, PCB, PCP, PBDE, and short-chain chlorinated paraffins.

TC 264 is linked with ISO/TC 146 and deals with ambient air emissions, indoor air workplace and other issues. This committee lists 112 working groups and subcommittees, most of which have been disbanded. SC6 is concerned with indoor air. TC 264 also has 45 working groups many of which are concerned with emissions, though mostly from ambient air.

Their website lists standards (EN 15267-1, 2 2023)
Air quality – Assessment of air quality monitoring equipment – Part 1: General principles of certification

17911: 2023
Stationary source emissions – Determination of total mercury – Automated measuring systems

17628: 2022
Fugitive and diffuse emissions of common concern to industry sectors – Standard method to determine diffuse emissions of volatile organic compounds into the atmosphere

16868:2019
Ambient air – Sampling and analysis of airborne pollen grains and fungal spores for networks related to allergy – Volumetric Hirst method

These and many other standards cost between €76 and 100 to purchase or you can just pay per day. Most are concerned with industrial problems and ambient air.

CEN/CENELEC European Guide on Standards and Regulation – Better Regulation through the use of voluntary standards – Guidance for policy makers. Edition 1 2015-6. **(113), (114)**

Table 4.6 CEN has 2163 active Technical Bodies

- **317** CEN Technical Committees
- **102** CEN Workshops
- **57** Sub-Committees of CEN Technical Committees
- **1519** CEN Working Groups
- **18** CEN-CENELEC Technical Committees
- **17** CEN-CENELEC Workshops
- **35** CEN-CENELEC Working Groups
- **3** CEN-CENELEC-ETSI Technical Committees
- **4** CEN-CENELEC

EU levels

"Levels" is a set of standards developed in Europe using "core sustainability indicators to measure carbon, materials, water, health, comfort and climate change impacts throughout a building's full life cycle." It appears to be a flexible and innovative solution to the complex problem of agreeing common standards across different European countries, but it is not at all clear whether this approach will be widely adopted, as debate continues about various other approaches such as the Construction Products Regulations (CPR) and the Energy Performance of Buildings Directive (EPBD). **(115)**

It's not clear how much "Levels" corresponds with other CEN and ISO standards.

Much of the work on "Levels" has been headed by Nicholas Dodd who has spent some time working at the Rocky Mountain Institute in the USA.

> Developed as a common EU framework of core indicators for assessing the sustainability of office and residential buildings, Level(s) can be applied from the very earliest stages of conceptual design through to the projected end of life of the building. As well as environmental performance, which is the main focus, it also enables other important related performance aspects to be assessed using indicators and tools for health and comfort, life cycle cost and potential future risks to performance. Level(s) aims to provide a common language of sustainability for buildings. This common language should enable actions to be taken at building level that can make a clear contribution to broader European environmental policy objectives. **(116)**

EU sustainable construction standards appear to have avoided the issue of indoor air quality and while there are some CEN standards on methods of measuring emissions, it is largely ignored throughout the many policies. However, Levels have tried to include IAQ.

Unfortunately, this document is very limited in its approach, setting out measurements for ventilation rate, CO2, relative humidity, TVOCs, benzene, radon, particulates and formaldehyde, but this is a big step forward in terms of EU policy.

> In terms of limiting VOC emissions at source, users should focus efforts on the following construction products and materials:
>
> Ceiling tiles, Paints and varnishes, including those applied to stairs, doors and windows, Textile floor and wall coverings, Laminate and flexible floor coverings, Wooden floor

> coverings, associated adhesives and sealants. Internal insulation products and any special interior surface treatments (e.g. damp-proof courses) shall be included within the scope too. **(117)**

Cherry picking a handful of emissions sources indicates a somewhat limited approach to IAQ, which is confirmed by the strange inclusion of a proposal to plant fir trees to deal with outdoor air pollutants.

> The intake of outdoor air pollutants (e.g. fine dust and benzene) will be minimised by placing the ground level intakes on the side of the building that is exposed to the car-park and not the main road and by the sheltering of ground-level air intakes by a row of densely planted fir trees. **(118)**

The main problem with Levels is the assumption that IAQ problems are a result of internal finishing materials and internal insulation materials. They have not considered the fact that when IAQ emissions are measured, the source for many of these can come from the fabric of the building which can be behind finishing layers. At least insulation materials are seen as a potential problem and they do refer to source control and low-emission materials.

Levels does include a section on "How to Specify low VOC emissions materials," and they mention the issue of third-party testing and the need for a EU harmonised system. L.2.2 Step 6 includes a section about mould and damp, but there is much that is wrong with how this is dealt with. They fail to discuss the importance of hygroscopic materials which can regulate humidity and the document places too great a reliance on mechanical ventilation.

There are a small number of independent bodies such as ANEC **(119)** and ECOS **(120)**, who do their best to represent consumers and environmental groups and are able to support a limited number of representatives on CEN working groups and other committees. They have been involved in discussions about EU policy on IAQ.

The EU Energy Performance of Buildings Directive (EPBD4)

The EU Commission published proposals to recast the EPBD in December 2021. The proposals are still to be finalised with possible implementation in the coming year. This may not affect UK Government energy policy. Aspects of the draft include:

- a new 'zero-emissions building' standard;
- setting limit values for whole-life global warming potential for new buildings to apply from 2030;
- a ban on the use of heating systems using exclusively fossil fuels;
- indoor environmental quality measurements to be taken in new buildings and buildings which have undergone major renovation to at least include the level of carbon dioxide, the temperature and thermal comfort and the relative humidity. There do not seem to be any plans to limit hazardous emissions from materials.

Attempts to include IAQ standards in the Construction Products Regulations

An international conference was held on April 20/21, 2021: "Limiting Health Impacts of Construction Products Regarding VOC."

The aim of the conference was to revitalise the discussions on a harmonised communication about VOC emissions. A harmonised classification and declaration approach belong to essential elements of improving the protection level for humans and the environment and ensuring healthy buildings.

Construction products are an important source of VOC emissions into indoor air. A chamber test was standardised under the Construction Products Regulation to enable harmonised measurement results of VOC emissions from construction products. If required by the product standards, the VOC emissions must be reported as part of the performance declaration. To provide understandable information to the product users, a class system was proposed by some of the Member States to differentiate the VOC emission performance into five levels. According to the proposal, four performance characteristics are evaluated to allocate a product to a VOC emission class. The overall feedback to the proposal was positive.

Speakers at the April 20/21, 2021 conference included Lilian Busse, Derrick Crump, Alessandro D'Amico, Katleen de Brouwere, Manfred Fuchs, Martin Glöckner, Heidrun Hofmann, Andreas Kortenkamp, Thomas Lützkendorf, François Maupetit, Birgit Müller, Michael Neaves, Alexander Röder, Edmund Vankann, Anna Vikström, Axel Vorwerk, Pekka Vuorinen, Olaf Wilke and Andrea Klinge of ZRS architects. **(121), (122)**

Unfortunately, despite the high degree of consensus at this important conference, little progress has been made persuading the various European bodies to adopt a harmonised class system for VOC emissions. In the meantime, different countries have different approaches, or none at all.

Different standards across different EU countries

Several EU countries have introduced indoor air quality and emissions standards but they vary in format. Air quality and healthy building activists have been campaigning for harmonised standards, and set their sights on revisions to the construction product regulations which were recently up for discussion, but there has been little progress.

The French Agence Nationale Sécurité Sanitaire Alimentaire Nationale (ANSES) is a food safety agency but has also developed standards for indoor air pollutants since 2011. **(123)**

A dozen pollutants of interest in indoor air have been appraised by ANSES with regard to the IAQGs and their detailed table can be downloaded and is in English

Formaldehyde
Carbon monoxide
Benzene
Naphthalene
Trichloroethylene and the addendum to the report
Tetrachloroethylene and the addendum to the report
Particulate matter
Hydrocyanic acid
Nitrogen dioxide
Acrolein
Acetaldehyde
Ethylbenzene
Toluene **(124)**

A number of surveys and published reports on indoor air pollutants have been carried out in France. **(125), (126)**

Germany

The German Environment Agency (Umwelt Bundes Amt – UBA) has been working on indoor air quality for a number of years and has linked this with the German AGBB (Committee of Health Evaluation of Building Products). The UK has nothing similar to such an organisation. The AGBB carries out emissions tests of products in a test chamber but is frequently called to inspect buildings where occupants have complained of problems. Inspection of flooring in one study revealed levels of Napthalene that were three times the AGBB safe limit and the carcinogenic 1,3 Dichloropropanol was 22 times the safe limit. **(127)**

> The use of low emission construction products is a key measure to facilitate good indoor air quality. This has become an even more prominent issue in the context of modern construction, particularly in energy-efficient buildings where the rate of air exchange with fresh ambient air may be limited.
>
> Harmonization of health requirements for construction products is a common objective for all parties involved. From the German point of view, it is essential that the future VOC classes are compatible with a high level of protection for all European building users. Only safe (construction) products should be placed on the market. **(128)**

Belgium

Policy measures in Belgium have been introduced to minimise the health impact of VOC emissions. This includes both VOC source control measures, and on-demand monitoring of indoor air quality (VOCs) in indoor dwellings in case of health complaints that are suspected to be caused by bad indoor quality. Flanders is one of the first regions in Europe with legally embedded health-based guidance values for indoor air quality called the Flemish Indoor Decree which they plan to broaden. GPs and Environmental health officers are trained to recognise health problems that might be caused by indoor air quality problems and can carry out indoor air quality testing. Flanders has adopted a set of compulsory safe levels for emissions and with target guidance values for 23 substances, mainly VOCs, such as 260ug/m3 for styrene emissions. **(129)**

EU regulation of economic activity since 2021 now includes restrictions on carcinogenic materials in building including paints and varnishes, ceiling tiles, floor coverings, adhesives and sealants, internal insulation and surface treatments for mould and damp, though not wood composite materials. Section 7.5 on pollution prevention and control states:

> Building components and materials used in the construction (and building renovation) that may come into contact with occupiers emit less than 0,06 mg of formaldehyde per m3 of material or component upon testing in accordance with the conditions specified in Annex XVII to Regulation (EC) No 1907/2006 and less than 0,001 mg of other categories 1A and 1B carcinogenic volatile organic compounds per m3 of material or component, upon testing in accordance with CEN/EN 16516 (290) or ISO 16000-3:2011 or other equivalent standardised test conditions and determination methods. **(130)**

Ireland

The Irish Green Building Council introduced a Home Performance Index (HPI) in 2018 that was amended in 2022, and includes a set of requirements for indoor air quality, referred to in 2018 as voluntary, but it's not clear whether they are mandatory requirements in the later version. The HPI contains a reference to Low Specification and Testing to "ensure good indoor air quality and avoid negative impact on VOCs." This has to be based on a postconstruction test to measure the levels of formaldehyde and on an "evaluation of low-VOC paints." This was to be carried out in accordance with ISO 160000-4-2011 which is a method for determining formaldehyde using diffusive sampling techniques. ISO 160000 standards are largely for work in test chambers, not testing air quality in buildings, though it does lay down the basis of gas chromatography testing in laboratories. This appears to have been drafted by someone unfamiliar with the practicalities of doing postconstruction testing in buildings. The 2022 version offers guidance "for testing method" referring to a web link to a document (IEC 847:1988) about local area networks that has nothing to do with indoor air quality and is also a withdrawn document!

This version does include a requirement for IAQ monitors to be installed in a minimum of two locations within the home, but these are only intended to monitor temperature, humidity and CO2 levels (what is normally measured by so-called low-cost sensors). It is not clear why VOCs and formaldehyde are not included. **(131)**

No evidence could be found of any published data from such assessments of buildings that have been awarded the HPI certification. A new Healthy Homes Ireland has been launched in 2023 and is discussed in a later chapter.

The original 2018 index was drafted with the help of commercial companies, including Saint Gobain, Kingspan, Schneider electric, MKN Property Group and Castlethorn construction and rather strangely the Canadian embassy. No reference to the HPI was found on the Castlethorn Construction website of their recent housing schemes or on the MKN property group website. **(132), (133)**

Traffic and industrial pollution

Most of the research funded by the UK Government on clean air and pollution has been focussed on external ambient air with emissions from traffic and industrial sources, though in the past year focus has also shifted to domestic burning in open fires and stoves as outlined above. There is no doubt that traffic and other external pollution can be very bad for health and many people suffer from a range of health conditions. The death of Ella Kissi Debra and the ruling by the Coroner that her asthma was caused by traffic pollution, where she lived, near a very busy stretch of the South circular road in London, has had a big impact on concern about outdoor emissions. Aged nine, Ella died from an asthma attack. Her mother has gone on to campaign for clean air through a foundation named after Ella and legislation in parliament has been termed Ella's law. **(134)**

Despite greater awareness of traffic pollution, there is still much for the campaign to do to improve the response of the medical profession to illness brought on by exposure to particulates from pollution. However, public awareness of these issues is mixed with hostility toward clean air zones that have been introduced in a number of cities. It is still common to see drivers sitting in cars outside supermarkets and schools with their engines idling for no obvious reason, and while many people have bought electric cars, these are very expensive and not accessible to poorer people, who may need a car to get to work. Traffic pollution remains a problem but it is largely confined to areas near busy roads, though many traffic pollution campaigners present it as widespread throughout the UK, even in rural areas where there is little traffic. Particulates can be generated by industry and agriculture, and health problems as a result of ammonia pollution from farming are just as serious.

Industrial pollution does not receive as much attention as traffic pollution and there seems to be insufficient monitoring and control. In South Wales, speed limits on the M4 have been lowered to 50 miles an hour as part of a Welsh Government environmental scheme, but these are within close proximity to large factories such as the Port Talbot steel works and the Rockwool factory at Bridgend (which is going through a large extension). Efforts to find data on emissions from these factories has been difficult, whereas similar data could be obtained from the EPA in the USA for similar factories. **(135)**

A local community council (Coychurch Higher) raised concerns about a number of issues associated with the Rockwool factory. They wanted to know what chemicals were coming out of the stacks and whether it is monitored and whether SO2 levels and other gases were monitored. But why are there no alarms for Rockwool plumes falling on residents?

The council report stated that the environmental and air quality concerns, potential noise pollution and potential highways issues are addressed, though the issues raised by local people did not seem to be fully covered. However, in a report as part of a planning application for an extension to the factory, a tunnel for dormice was to be included under a conveyor on the site. The council report on a recent Rockwool planning application stated that Rockwool fibres pose no risk to human health if inhaled.

The UKRI clean air programme and external air pollution

The UK Government has been funding research into Clean Air through a consortium of bodies.

Wave 1 of this Clean Air Programme, with £42.5million of Government money, has largely focussed on external air and traffic pollution and a great deal of useful information can be obtained from these projects even though indoor air quality was not on the agenda. **(136)**

In Wave 2 of the programme, some projects have begun to look into indoor air quality problems though, as will be seen below, some rather curious assumptions were made about the main issues on which to focus. At the time of writing much of this work is ongoing and so it's not possible to report the conclusions.

The author was asked to chair a steering committee related to the Clean Air Programme. Its role seemed mainly to monitor progress and it did not take any decisions about funding and instead is now focussed on how to disseminate the results of research. Decisions about who should receive funding under the Clean Air Special Project fund had been made by a small group of officials from the various funding bodies, including UKRI, NERC, Met. office, Innovate UK, EPSRC ESRC, MRC and so on. Final decisions about who should receive grants was made by a small group of officials known as the programme board.

The research brief for Wave 2 talked about the interface between indoor and outdoor pollution. This was not necessarily a bad place to start as a significant part of bad indoor air quality can be caused by particulates and emissions coming into houses through open windows and ventilation systems, but it may have been a concession to those academics who remained pre-occupied with traffic pollution. Many of the projects that have been funded refer to the indoor/outdoor interface but this remains a largely unexplored concept from a scientific perspective.

A summary of the Wave 2 Clean Air projects

TRANSITION Clean Air Network, University of Birmingham, says it connects researchers across nine UK higher education institutions, Public Health England and more than 20 stakeholder partners spanning the private, public and civil sectors to address clean challenges associated with a low-emission transport and mobility revolution in the UK. **(137)**

The IAQ-EMS project is another project at the University of Birmingham with aims to develop software and data tools to advance the UK's capacity for *indoor air quality modelling*, for estimation of emissions and exposure. Collaboration within this project will draw on the expertise of investigators which includes the areas of air quality monitoring, chemical and dynamical modelling, built environment and public health. This project focuses on the use of "low-cost sensors," studying indoor particulate matter from cooking in UK students' studio flats and evaluation of cooking methods, PM concentrations and personal exposures using low-cost sensors. **(138)**

This team claims that cooking emissions have been identified as a major source of indoor PM, which can contribute to severe health issues, including cardiovascular disease and lung cancer, though there is no explanation of how cooking has been identified and by whom. It will be interesting to see what levels of PM are detected in student flats, as this project suggests that cooking water-based dishes while operating extractors would improve indoor air quality and reduce PM exposure. Other indoor air pollutants such as from building materials do not seem to have been considered. CleanairforV is another project at the University of Birmingham meant to identify, develop and evaluate indoor air pollution solutions for two vulnerable groups (VGs): children (aged 0-16) and people with pre-existing conditions (e.g., COPD). The project seems largely to be about building a network rather than carrying out empirical research, but a diagram of one of their work packages is about external pollution based on particulate exposure from PM1, ultrafine PM and VOCs at the" indoor/outdoor interface" in nurseries, schools hospitals and homes and transport hotspots. A search of their website did not provide any details on what VOCS were to be measured or how this would be done. **(139)**

Engaging the public: Indoor and Outdoor Air Quality research is being carried out by Global Action Plan, partnered with the Met Office. This is essentially an education and communication project. They say that they will design and demonstrate science-based materials with and for children and young people to help them better understand and act on air pollution and groups of adults with different needs and circumstances with respect to air pollution knowledge. The say that the project will build on the UK's air quality research and be guided by a panel of academics to ensure research findings are incorporated into the materials and they have created a Clean Air Hub. While indoor air quality gets a mention in the materials already produced by GAP it was impossible to find anything about indoor air quality in the hub. GAP's approach puts the emphasis on people taking responsibility for their health and wellbeing. **(140), (141)**

Bioarnet is a project at Cranfield University. It is a network of scientists concerned primarily with external air pollution in the form of particulates. They say that Public Health England concluded that, in the UK, between 28,000 and 36,000 deaths a year can be attributed to long-term exposure to particulate pollution, but they say that they still do not know what role PM of biological origins or bioaerosols play in this health burden nor the precise impact of exposures within the "*indoor or outdoor continuum*." Maybe this project will provide more solid evidence of this concept, but it is not clear how this project will collect and analyse this data. The network brings together 200 academic, business and other community-based stakeholders through a range events and conferences creating something they call a "bioexposome." **(142), (143)**

The HEICCAM network at the University of Edinburgh also refers to the indoor/outdoor interface. This project appears to overlap with other clean air programme projects, with most of its outputs being from other projects such as Breathing City and TAPAS (see below). They have published a series of YouTube presentations they say will create an understanding of key trade-offs and future challenges. It took some searching to find any reference to indoor air pollution which they list as from cooking, fabrics and furnishings, cleaning and personal care products. HEICCAM submitted evidence to parliament in May 2023 which refers to indoor air 65 times but simply stating that indoor air "is a problem." They promote the use of low-cost sensors but without any reference to emissions from building materials.

> Large-scale monitoring using low-cost sensors (which are becoming increasingly available and affordable) is key to examining different buildings and how indoor air quality varies in multiple scenarios. For example, labelling of sources of indoor air pollutants (such as VOC emitters) and alerts to their potential health effects may influence occupant behaviour. **(144), (145)**

The TAPAS project at University of Cambridge claims to have a team of 18 people including academics from other Clean Air-funded projects concerned with air pollution in schools. They are collaborating with Professor Prashant Kumar, founding director of the Global Centre for Clean Air Research (GCARE), and say that "The best way to limit the exposure to children is to control the air pollutant emissions at the source." However, it seems that this is referring to external pollution. Exploring the TAPAS website leads to other projects funded under the Clean Air programme. Several of these projects appear to have a great deal of overlap. They advocate the use of CO2 sensors in schools. **(146), (147)**

The Immaterial project at Cambridge University claims that a top priority in the UK Clean Air strategy is to remove NO2. They are working with York University who will be testing "novel porous materials" that can capture NO2 pollution. NO2 pollution comes from combustion petrol fuel and gas cooking. **(148), (149)**

The Breathing City project at Leeds University is another network organisation which plans to develop a "holistic" methodology to manage exposure to air pollution by considering the airflows between indoor and outdoor environments. Their website takes us to another website about "future urban ventilation." Based on work by the National Engineering Policy Centre **(150),** this work was commissioned by the Government Chief Scientific Adviser Sir Patrick Vallance to identify the interventions needed in the UK's built environment and transport systems to reduce infection transmission. Another link on the Breathing city website takes us to "Theme 1 coupled" indoor-outdoor environment presentations by Cook and Reeuwijk, which presents us with technical requirements for mechanical ventilation modelling, largely related to large office buildings. One of the presentations as part of this work does provide some interesting work on measuring CO2 emissions in buildings by Dr. Liora Malki Epstein from University College London. This project appears to overlap with some of the other funded projects. **(151)**

The Respire project at Swansea University is concerned with the ill effects of air pollution on pregnant women and is focussed on "airborne materials," and how air pollution passes to the baby possibly leading to poor health in childhood. This is being done by taking biological samples from pregnant volunteers at various trimesters including nasal samples, peripheral and umbilical cord blood, placenta and sperm.

> Samples will be exposed to PM2.5 or fine particulate matter, a cocktail of chemical and biological contaminants including house dust and volatile organic compounds, such as the chemicals found in cleaning products, alone and in combination including with other airborne materials such as pollen and viruses.

An important aspect of this study is that they are measuring natural exposures in the homes of pregnant women, how the woman is responding to this environment and then follow the health of the baby as it grows up. It will be interesting to see whether the samples obtained show emissions from endocrine-disrupting chemicals such as flame retardants and carcinogens such as formaldehyde, which of course come from building materials, but they say they are exposing the samples to their own chosen VOCs from cleaning products rather than other VOC sources apparently. **(152)**

The Codikoat project is run by a commercial private company, CodiKoat, which produces antimicrobial and antiviral coating materials which they claim kills viruses, bacteria and fungi. These coatings are being used in extract fans and HEPA filters. The "Codivent" filter materials are meant to remove "harmful small particles and poisonous gasses" which they claim emanate from cooking. The project says it will test the efficacy of this, but it was not clear if there were any independent organisations that would be carrying out such evaluation. **(153), (154)**

The ARBNCO project is a private business that says it will monitor, visualise and reduce the health impacts of domestic air pollution. However, their website suggests their work is largely about decarbonisation (reducing energy consumption) and it was hard to find anything about indoor air pollution. They seem to promote sensors that give TVOC. The work also involves a "Living Lab" which says is measuring pollution in homes and will create a technology that people can use in their homes. Apparently ARBNCO have won awards for their "Well platform" which collects data on temperature and TVOCs. The only award that could be found was for project innovation from CIBSE. They do say that they will offer recommendations to householders to mitigate exposure to pollutants, but as they are not collecting data from real emissions such as flame retardants, formaldehyde and a host of other VOCS it is hard to work out what this might involve. The University of Strathclyde is involved so there may be some independent evaluation. **(155), (156)**

FamilyAir is another private business led by Filament PD Ltd. This project aims to develop a "cost effective, human centred air quality monitoring system" that engages with each member of a family personally, helping them understand and improve air quality within their home" as they say that people do not have the technical knowledge to deal with more sophisticated equipment. If people are alerted to bad air quality through such as system...what do they then do about it? Nothing is suggested. Several attempts were made to contact Family Air but there was no one in their office to answer phone calls and there has not been any response to messages., When contacted, the person answering the phone at Filament PD said they had not heard of the Family Air project, but they have received £300,000 funding. No independent evaluation of this project was listed. **(157)**

Smarter Home Indoor air quality monitoring system is another private company project run by Applied Nano-detectors Ltd. This is similar to Family Air in planning to develop an "ultrasensitive home-based IAQ sensing system" that can detect adverse pollution levels, predict and identify the pollution sources and provide suggestions to help people improve IAQ. The new monitor will apparently be field-tested in households across the UK. Using this new system will enable users to correlate household tasks to pollution events and take action to reduce them it is claimed, and that this will lead to behaviour change and positive action to reduce IAQ that affects their health. There is the implication that bad indoor air quality is a result of human behaviour.

The system will measure Carbon Dioxide (CO2), Particulate Matter (PM 1.0, PM 2.5 and PM 4.0), Total Volatile Organic Compounds (TVOC), Formaldehyde, Humidity and Temperature. TVOC information will be of limited value but at least Formaldehyde can be detected. Householders will receive a picture using "augmented reality visualisation" on their mobile phone saying that their indoor air is good or bad. This project states that indoor sources of pollution stems from combustion sources, namely "tobacco, cleaning supplies, paints, and insecticides wood, coal heating, cooking appliances, and fireplaces which release harmful by-products such as carbon monoxide directly into the indoor environment." Strangely Carbon Monoxide detection is not mentioned as a feature of their detection. It was not possible to find if anyone is involved in carrying out an independent evaluation. Applied nano-detectors were contacted and asked how many houses were taking part in the field testing but no reply has been received. **(158)**

The Consortia project at Manchester University says it will achieve a fundamental understanding of toxicological mechanisms causing adverse health outcomes from exposure to various air pollutants. They have built a chamber that they will use to simulate different types of air pollution exposure, but they have only selected diesel exhaust fumes, wood smoke and cooking emissions for this project. Emissions from buildings materials are not being considered.

On a visit to the chamber, a large VW diesel engine was parked in the laboratory to be used with the chamber. Their aim is to see what impact they call "common pollutants" has on neurological disease and cognitive function. It is open to question whether they have selected "common pollutants," but there seems little doubt that the volunteers who will enter the chamber are not going to feel well if it fills with diesel exhaust fumes! Apparently, there was a lot of negotiation with the ethics team at Manchester Royal Infirmary because real live volunteers will be used. It will be interesting to see what meals are cooked to test cooking. The controversial issue of wood smoke is discussed above. **(159)**

The Ingenious Project at the University of York involves three other universities and Bradford Teaching Hospital and the Born in Bradford study. They say they will track and quantify the composition of pollutants within indoor spaces, such as those from cooking and cleaning, how air pollutants from indoor air sources affect outdoor air quality and vice versa and how different occupant behaviours affect production of and exposure to air pollutants. While there is discussion of the interface between internal and external pollution it hardly seems worth wasting valuable scientific time investigating how indoor air pollution might affect external air.

The project will use the Born in Bradford project to quantify and identify air pollutants, how air pollutants react chemically and transform over time, including when different air pollutants mix, and how different household behaviours affect the production of and exposure to air pollutants. This will be done by installing sensors in 300 homes in Bradford to monitor cooking and cleaning and assess the effect on health. So yet again another study that has already decided the main problems are cooking and cleaning and the really bad emissions in the homes may be ignored, with occupant behaviour getting the blame, not the buildings. **(160)**

The Wellhome project at Imperial College London is looking at how indoor pollutants in nearby West London impact children with asthma. This four-year project, costing £9 million, is being carried out by a multi-disciplinary team who have involved local people in managing three linked research projects, and where they are monitoring 100 local households. Wellhome state that links between indoor air quality and public health is an under-researched area. The project will assess chemical and biological pollutants and monitor individual activity and health data. They have engaged with the nearby community and are linking up with groups like Mums for Lungs on Clean Air Day; it is worth reviewing their online news pages. **(161), (162)**

Will the Clean Air Programme throw new light on indoor air quality?

Overall the range of projects, funded under Wave 2 of the Clean Air Programme, are disappointing from the perspective of investigating indoor air quality. Emerging from the preoccupation with external air pollution an opportunity has been missed here to investigate indoor emissions, except possibly the project at Imperial College and maybe the Bradford study. As the Wellhome projects states, poor air quality is a severe environmental risk to public health in the UK, responsible for 40,000 early deaths, and it is estimated this costs £20 billion a year to health services and businesses.

It was surprising to find how many projects have begun by making the assumption that indoor air problems are due to cooking and cleaning and even diesel emissions. The assumption that cooking and cleaning are the main indoor air problems suggests that the scientists involved in these projects have not carried out a proper literature review before drawing up their proposals. Several projects have assumed that so-called low-cost sensors will provide useful data, which is open to question. It is also worrying that a number of private businesses have received funding without obvious independent evaluation of their work, but it may simply mean that this has not been made obvious. There is little doubt that the officials in charge of the Clean Air programme will have an overview of these projects and will ensure that they are carried out properly. At the time of writing there is no indication that any further UK Government funding will be available to support proper research work on indoor air quality. As chair of the Clean Air Steering Committee, it was too late to influence the selection of these projects, but there may be opportunities to call for further funding on indoor air quality research in the future, particularly when it becomes clear that they have failed to address the real causes of bad indoor air quality.

A recent encouraging development is the publication of a Post Note on Indoor Air Quality. This document is part of a series of briefing notes for members of the Westminster parliament. Unlike the DEFRA, AQEG and NICE reports it pays more attention to the issue of emissions from building materials though sadly still devoting too much space to wood burning stoves and cooking. **(163)**

Last minute news before going to press with this book is that the TCPA Healthy Homes bill is still making its way through Parliament and appears to have been incorporated into a Levelling Up and Regeneration bill which includes reference to Healthy Homes and includes Schedule 7 which, under the heading of biodiversity, incorporates the principles set out in the TCPA bill. While it is doubtful that this will have much effect on new private or social housing developments, as the principles are so vague and general, the relevant responsible authorities for implementing the legislation may be persuaded to take more detailed measures onboard. They are local planning authorities; public health authorities; urban development corporations; new town development authorities; the planning inspectorate and Homes England.

References

(1) Hodkinson, S. (2019). *Safe as Houses*. Manchester: Manchester University Press.
(2) Grenfell United [RSH 108]. (2021). *Executive summary*. Available at: https://committees.parliament.uk/writtenevidence/42211/html/#:~:text=%5B2%5D%20The%20Inquiry's%20terms%20of,White%20Paper%20in%20November%202020 [Accessed 19 Sep. 2023].
(3) Apps, P. (2022). *Show Me the Bodies*. London: Oneworld Publications.
(4) HERITAGE HOUSE (n.d). Building Regs Review – Hackitt conflict. Available at: www.heritage-house.org/stuff-about-old-buildings/insulation/building-regs-review-hackett-conflict.html [Accessed 19 Sep. 2023].
(5) Dempsey, H. (2021). 'Europe's chemicals industry warns on threat from EU plans.' *Financial Times*. Available at: https://www.ft.com/content/4aa2d3e3-5888-4952-8349-aad0913b3ef3 [Accessed 19 Sep. 2023].
(6) Energy Saving Trust (n.d). Independent verification. Available at: https://energysavingtrust.org.uk/business/energy-efficiency/verified/ [Accessed 19 Sep. 2023]

(7) Energy Saving Trust. (2020). Q-bot. Available at: https://energysavingtrust.org.uk/case-study/q-bot/ [Accessed 19 Sep. 2023]
(8) WIKIPEDIA. (2023). *Judith Hackitt*. Available at: https://en.wikipedia.org/wiki/Judith_Hackitt [Accessed 19 Sep. 2023].
(9) GOV.UK. (2017). Expert panel appointed to advise on immediate safety action following Grenfell fire. Available at: www.gov.uk/government/news/expert-panel-appointed-to-advise-on-immediate-safety-action-following-grenfell-fire [Accessed 19 Sep. 2023].
(10) FSM. (2022). Key Grenfell witness 'must be held accountable' for his actions urges FBU. Available at: www.fsmatters.com/Government-advisors-must-be-held-to-account#:~:text=In%20the%20years%20prior%20to,Brigade%20from%202003%20until%202007 [Accessed Feb. 24 2023].
(11) Robinson-Pasha, L. (2017). 'Grenfell Tower: Fire safety chief accused of 'conflict of interest' over cladding certification role.' *Independent*. Available at: www.independent.co.uk/news/uk/home-news/grenfell-tower-fire-safety-chief-sir-ken-knight-cladding-certification-conflict-interest-kensington-chelsea-council-london-a7876516.html [Accessed 19 Sep. 2023].
(12) Barratt, L. (2018). 'Company connected to government fire expert advised on Grenfell refurbishment.' *Inside Housing*. Available at: www.insidehousing.co.uk/news/news/company-connected-to-government-fire-expert-advised-on-grenfell-refurbishment-56608 [Accessed 19 Sep. 2023].
(13) HSE. (n.d). Building Safety Regulator. Available at: www.hse.gov.uk/building-safety/regulator.htm [Accessed 19 Sep. 2023].
(14) HM Government. (2022). Air Quality Common Framework: Provisional Framework Outline Agreement and Concordat. [pdf]. Available at: https://assets.publishing.service.gov.uk/government/uploads/system/uploads/attachment_data/file/1052059/air-quality-provisional-common-framework.pdf [Accessed 19 Sep. 2023].
(15) AQEG. (2022). Report: Indoor Air Quality. Available at: https://uk-air.defra.gov.uk/research/aqeg/publications [Accessed 19 Sep. 2023].
(16) Available at: https://environment.ec.europa.eu/topics/circular-economy/eu-ecolabel-home
(17) GEV-EMICODE. (n.d). Available at: www.emicode.com/index.php?id=1&L=1 [Accessed 19 Sep. 2023].
(18) GUT. (n.d). The new Product Passport for carpets and rugs. Available at: http://pro-dis.info/86.html?&L=0 [Accessed 19 Sep. 2023].
(19) Blue Angel. (n.d). Available at: www.blauer-engel.de/en/index.php [Accessed 19 Sep. 2023]
(20) Austrian Ecolabel. (n.d). Available at: www.umweltzeichen.at/cms/home233/content.html [Accessed 19 Sep. 2023].
(21) Umwelt Bundesamt. (n.d) Available at: www.umweltbundesamt.de/themen/gesundheit/kommissionenarbeitsgruppen/ausschuss-zur-gesundheitlichen-bewertung-von [Accessed 19 Sep. 2023].
(22) Available at: https://ymparisto.rakennustieto.fi/en/emission-classification-of-building-materials
(23) Available at: www.anses.fr/en
(24) Europur. (n.d). Available at: www.europur.com/index.php?page=certipur [Accessed 19 Sep. 2023].
(25) Danish Indoor Climate Label www.ecolabelindex.com/ecolabel/danish-indoor-climate-label
(26) Swedish Building Product Assessment www.byggvarubedomningen.se/
(27) Nature Plus. (n.d). Natural sustainable buildings. Available at: www.natureplus.org/ [Accessed 19 Sep. 2023].
(28) Algrim, L. B., Pagonis, D., de Gouw, J. A., Jimenez, J. L. and Ziemann, P. J. (2020). 'Measurements and modelling of absorptive partitioning of volatile organic compounds to painted surfaces.' *Indoor Air*, 30(4), 745–756. https://doi.org/10.1111/ina.12654 [Accessed 19 Sep. 2023].
(29) AQEG op cit.
(30) Lewis, A. C. (2018). 'The changing face of urban air pollution: Volatile organic compounds in U.S. urban air increasingly derive from consumer products.' *Science*. 359(6377), 744–745 Available at: https://doi.org/10.1126/science.aar4925 [Accessed 19 Sep. 2023].
(31) Lewis, A. C., Jenkins, D., & Whitty, C. J. M. (2023). Hidden harms of indoor air pollution – five steps to expose them. *Nature*, 614(7947), pp.220–223. Available at: https://doi.org/10.1038/d41586-023-00287-8 [Accessed 19 Sep. 2023].

(32) Building Research Establishment (BRE). (2022). BRE's research and expertise supports Defra's indoor air quality report. Available at: https://bregroup.com/news/bres-research-and-expertise-supports-defras-indoor-air-quality-report/ [Accessed 19 Sep. 2023] BRE. (n.d) *Indoor Environment Testing.* Available at: https://bregroup.com/services/testing-certification-verification/indoor-environment-testing/ [Accessed 19 Sep. 2023].

(33) Bone, A. et al. (2010). 'Will drivers for home energy efficiency harm occupant health?' *Perspectives in Public Health,* 130(5), pp.233–238. Available at: https://doi.org/10.1177/1757913910369092 [Accessed 19 Sep. 2023].

(34) Redmore Environmental. (n.d). BREEAM Indoor Air Quality. Available at: https://red-env.co.uk/services/breeam-indoor-air-quality-assessment/?gclid=EAIaIQobChMIvqO91pmigQMVFsXtCh0wjAw0EAAYASAAEgK1aPD_BwE [Accessed 19 Sep. 2023].

(35) BREEAM. (n.d). Hea 02 – Indoor Air Quality. Available at: https://kb.breeam.com/section/int-nc-2013/02-health-and-wellbeing-int-nc-2013/hea-02-indoor-air-quality-02-health-and-wellbeing-int-nc-2013/ [Accessed 19 Sep. 2023] National Institute for Health and Care Excellence (NICE). (2020). Indoor Air Quality at home. Available at: www.nice.org.uk/guidance/ng149 [Accessed 19 Sep. 2023].

(36) Nature Plus Institute. (n.d). Award criteria. Available at: https://natureplus-institute.eu/?page_id=46&lang=en [Accessed 19 Sep. 2023].

(37) Ecolabel Index. (n.d). Natureplus. Available at: www.ecolabelindex.com/ecolabel/natureplus#:~:text=The%20natureplus%C2%AE%2Dseal%20of,sustainable%20sources%20of%20raw%20materials [Accessed 19 Sep. 2023].

(38) National Institute for Health and Care Excellence. (2020). Indoor air quality at home: [3.1] Evidence review for material and structural interventions. [pdf] Available at: www.nice.org.uk/guidance/ng149/evidence/3.1-material-and-structural-interventions-pdf-7020943888 [Accessed 19 Sep. 2023].

(39) National Institute for Health and Care Excellence. (n.d) Guideline scope: Indoor air quality at home. [pdf] Available at: www.nice.org.uk/guidance/ng149/documents/final-scope [Accessed 19 Sep. 2023].

(40) National Institute for Health and Care Excellence Indoor Air Quality at Home. (2019). (2) Evidence review for exposure to pollutants and health outcomes 1.6.1

(41) NICE Recommendations Regulators and building control 1.6.1. www.nice.org.uk/guidance/ng149/chapter/Recommendations#regulators-and-building-control-teams [Accessed 24 Sep. 2023].

(42) All Party Parliamentary Group on Healthy Homes and Buildings. (n.d). Available at: https://healthyhomesbuildings.org.uk/ [Accessed 19 Sep. 2023].

(43) UK Parliament. (2022). 'Town and Country Planning Association Version 5 (England)' Available at: https://bills.parliament.uk/bills/3139 [Accessed 19 Sep. 2023].

(44) TCPA. (2023). Report Stage of the Levelling Up and Regeneration Bill: A briefing by the TCPA. [pdf] Available at: www.tcpa.org.uk/wp-content/uploads/2023/08/Lords-briefing-for-LURB-report-stage_September-2023-2.pdf [Accessed 19 Sep. 2023].

(45) The Ella Roberta Foundation. (n.d). Inquest & Coroner's Report. Available at: https://ellaroberta.org/campaigns/inquest-coroners-report#:~:text=Inquest%3A-,In%20what%20is%20believed%20to%20be%20a%20global%20first%2C%20Deputy,direct%20result%20of%20air%20pollution [Accessed 19 Sep. 2023].

(46) UK Parliament. (2023). Clean Air (Human Rights) Bill [HL]. Available at: https://bills.parliament.uk/bills/3161 [Accessed 19 Sep. 2023].

(47) Birkett, S. (2012). 'Guide: Indoor air quality can be worse than outdoor.' *Clean Air in London.* Available at: https://cleanair.london/hot-topics/indoor-air-quality-can-be-worse-than-outdoor/ [Accessed 19 Sep. 2023].

(48) CAMFIL. (n.d). Available at: www.camfil.com/en-gb/ [Accessed 19 Sep. 2023].

(49) Mulholland, H. (2011). 'Former banker's pollution fight a breath of fresh air.' *The Guardian.* Available at: www.theguardian.com/environment/2011/feb/22/banker-air-pollution-fight#:~:text=Simon%20Birkett%20is%20fighting,%2Dups%20in%20modern%20history%22 [Accessed 20 Sep. 2023].

(50) www.gov.uk/guidance/february-2023-update-on-governments-work-to-improve-the-quality-of-social-housing#:~:text=The%20Bill%20will%20also%20strengthen,with%20landlords%20footing%20the%20bill.

(51) Hearne, J. (2020). 'Hell's kitchen – Why cooking can destroy indoor air quality.' *Passive House.* Available at: https://passivehouseplus.co.uk/magazine/insight/hell-s-kitchen-why-cooking-can-destroy-indoor-air-quality [Accessed 20 Sep. 2023].

(52) World Health Organisation (WHO). (2022). Household air pollution. Available at: www.who.int/news-room/fact-sheets/detail/household-air-pollution-and-health [Accessed 20 Sep. 2023].

(53) GOV. UK. (2022). Carbon monoxide: general information. Available at: www.gov.uk/government/publications/carbon-monoxide-properties-incident-management-and-toxicology/carbon-monoxide-general-information [Accessed 20 Sep. 2023].

(54) McMahon, J., Unkel, C. and Lein, P. J. (2023). 'Out of the frying pan and into the fire: The gas stove toxicity debate.' *Open Access Government.* Available at: https://doi.org/10.56367/OAG-038-9559 [Accessed 20 Sep. 2023].

(55) RMI. (n.d). Health and air quality impacts of cooking with gas. Available at: https://rmi.org/press-release/health-air-quality-impacts-of-cooking-with-gas/#:~:text=Gas%20stoves%20release%20several%20hazardous,levels%2C%20can%20cause%20respiratory%20effects [Accessed 20 Sep. 2023].

(56) NAQTS. (n.d). Indoor and outdoor air quality monitoring. Available at: www.naqts.com/ [Accessed 20 Sep. 2023].

(57) Kearney, N. and Blair, H. (2023). 'UK policies do not protect the public from the impacts of gas cooking.' *CLASP.* Available at: www.clasp.ngo/updates/uk-policies-do-not-to-protect-the-public-from-the-impacts-of-gas-cooking/ [Accessed 20 Sep. 2023].

(58) Blair, H., Kearney, N. and Scholand, M. (2023). 'Exposing the hidden health impacts of cooking with gas.' *CLASP.* Available at: www.clasp.ngo/research/all/eu-gas-cooking-health/ [Accessed 21 Sep. 2023].

(59) Clasp. (n.d). Electrifying cooking in Europe. Available at: www.clasp.ngo/cook-cleaner-europe/ [Accessed 21 Sep. 2023].

(60) Modern Energy Cooking Services. (n.d). Available at: https://mecs.org.uk/ [Accessed 21 Sep. 2023].

(61) Mathiesen, K., Bencharif, S. T. and Zimmerman, A. (2022). 'Gas stoves might be a killer in the kitchen.' *Politico.* Available at: www.politico.eu/article/gas-stove-killer-health-climate-pollution/ [Accessed 21 Sep. 2023].

(62) Lim, Y. H. (2022). 'Inflammatory markers and lung function in relation to indoor and ambient air pollution.' *International Journal of Hygiene and Environmental Health*, 241. Available at: https://doi.org/10.1016/j.ijheh.2022.113944 [Accessed 21 Sep. 2023].

(63) Wilkins, B. (n.d). 'Industry knew–and hid–dangers of gas stoves over 50 years ago.' *Common Dreams.* Available at: www.commondreams.org/news/gas-stoves [Accessed 21 Sep. 2023].

(64) Lawrence, G. H. (1972). Energy and the environment…a crisis: natural gas…a solution. [pdf] Available at: www.documentcloud.org/documents/23690657-197216_energy_and_the_environment_a_crisis_aga_draft_report_plus_cover_letter_to_nipcc_exec_director [Accessed 21 Sep. 2023].

(65) Davis, N. (2023). 'Scientists find oldest known evidence of humans in Europe using fires to cook.' *The Guardian.* Available at: www.theguardian.com/science/2023/may/18/scientists-find-oldest-known-evidence-of-humans-in-europe-using-fires-to-cook [Accessed 21 Sep. 2023].

(66) Heydon, J. and Chakraborty, R. (2022). 'Wood burning stoves, participatory sensing, and 'cold, stark data'' *SN Social Science*, 2, 219. Available at: https://doi.org/10.1007/s43545-022-00525-2 [Accessed 21 Sep. 2023].

(67) DEFRA. (2020) *Estimating UK domestic solid fuel consumption, using Kantar data: Summary of results and discussion.* [pdf] Available at: https://cleanair.london/app/uploads/14973_AnnexeA-UKdomesticsolidfuelsuseestimatespaper.pdf [Accessed 21 Sep. 2023].

(68) KANTAR. (n.d). Data Solutions. Available at: www.kantar.com/expertise/research-services/data-solutions [Accessed 21 Sep. 2023].

(69) Clean Air Programme. (2023). Domestic Burning Workshop. Available at: www.ukcleanair.org/2023/06/16/domestic-burning-workshop-cardiff-28-june-2023/ [Accessed 21 Sep. 2023].
(70) Carrington, D. (2022). 'Wood burners emit more particle pollution than traffic, UK data shows.' *The Guardian.* Available at: www.theguardian.com/environment/2022/feb/15/wood-burners-emit-more-particle-pollution-than-traffic-uk-data-shows [Accessed 21 Sep. 2023].
(71) Harvey, F. (2023). 'Rise in UK wood-burners likely to be creating pollution hotspots in affluent areas.' *The Guardian.* Available at: www.theguardian.com/environment/2023/feb/06/rise-in-wood-burners-in-uk-likely-creating-new-pollution-hotspots-in-affluent-areas [Accessed 21 Sep. 2023].
(72) Carrington, D. (2021) 'Eco wood stoves emit 750 times more pollution than an HGV, study shows.' *The Guardian.* Available at: www.theguardian.com/environment/2021/oct/09/eco-wood-stoves-emit-pollution-hgv-ecodesign [Accessed 21 Sep. 2023].
(73) Trechera, P. et al. (2023). 'Phenomenology of ultrafine particle concentrations and size distribution across urban Europe.' *Environment International*, 172, pp.107744–107744. Available at: https://doi.org/10.1016/j.envint.2023.107744 [Accessed 21 Sep. 2023].
(74) Rabajczyk, A., Zielecka, M. and Małozięć, D. (2020). 'Hazards Resulting from the Burning Wood Impregnated with Selected Chemical Compounds.' *Applied Sciences*, 10(17), p.6093. Available at: https://doi.org/10.3390/app10176093 [Accessed 21 Sep. 2023].
(75) HETAS. (n.d). Available at: www.hetas.co.uk/ [Accessed 21 Sep. 2023].
(76) Heydon and Chakraborty. (2022). Op cit.
(77) Byrne, R. et al. (2023). 'Highly local sources and large spatial variations in $PM_{2.5}$ across a city: evidence from a city-wide sensor network in Cork, Ireland.' *Environmental Science: Atmospheres*, 3, 919–930.
(78) 'On wood smoke & 'The Alchemist' – a poem' *gaia holmes.* Available at: https://gaiaholmes.wordpress.com/on-wood-smoke-the-alchemist-a-poem/ [Accessed 21 Sep. 2023].
(79) Dimitroulopoulou, S. and Shrubsole, C. (2020). 'Indoor Air Quality Guidelines for selected VOCs in the UK.' [pdf] *PHE.* Available at: https://asbp.org.uk/wp-content/uploads/2020/03/Sani-Dimitroulopoulou-Public-Health-England-ASBP-Healthy-Buildings-2020.pdf [Accessed 21 Sep. 2023].
(80) Shrubsole, C. et al. (2019). 'IAQ guidelines for selected volatile organic compounds (VOCs) in the UK.' *Building and Environment*, 165, 106382. Available at: https://doi.org/10.1016/j.buildenv.2019.106382 [Accessed 22 Sep. 2023].
(81) Department Of Finance. (2023). Consultation on a review of energy efficiency requirements and related areas of Building Regulations. Available at: www.finance-ni.gov.uk/consultations/review-energy-efficiency-building-regulations [Accessed 22 Sep. 2023].
(82) IQAIR. (2021). Indoor carbon dioxide. Available at: www.iqair.com/us/newsroom/indoor-carbon-dioxide-co2 [Accessed 22 Sep. 2023].
(83) Passivhaus Trust. (2020). Demystifying Airtightness. [pdf]. Available at: www.passivhaustrust.org.uk/UserFiles/File/Technical%20Papers/Good%20Practice%20Guide%20to%20Airtightness%20v10.6-compressed(1).pdf [Accessed 22 Sep. 2023].
(84) LETI. (2020). Climate Emergency Design Guide. Available at: www.leti.london [Accessed 22 Sep. 2023].
(85) Awbi, H. B. (2015). 'Indoor air quality in UK homes and its impact on health.' BEAMA. www.beama.org.uk/static/uploaded/22bd9b55-b137-4e43-b3a67e1bc1b08f78.pdf [Accessed 24 March 23].
(86) Making SAP and RdSAP 11 fit for Net Zero. (n.d). [pdf]. Available at: www.levittbernstein.co.uk/site/assets/files/3670/making_sap_and_rdsap_11_fit_for_net_zero-full_report.pdf [Accessed 22 Sep. 2023].
(87) Housing Ombudsman Service. (2023). 'Two severe maladministration findings for Clarion as Ombudsman reiterates recommendations from its special investigation.' Available at: www.housing-ombudsman.org.uk/2023/01/31/multiple-severe-maladministration-findings-for-clarion-as-ombudsman-reiterates-recommendations-from-its-special-investigation/ [Accessed 22 Sep. 2023].
(88) Levitt Bernstein. (n.d). Urban Design, Landscape Architecture Housing: Eastfields Estate, Merton. Available at: www.levittbernstein.co.uk/project-stories/eastfields/ [Accessed 22 Sep. 2023].
(89) Etude. (n.d). Available at: https://etude.co.uk/about/ [Accessed 22 Sep. 2023].

(90) Ministry of Housing, Communities & Local Government. (2019). Ventilation and indoor air quality in new homes. Available at: www.gov.uk/government/publications/ventilation-and-indoor-air-quality-in-new-homes [Accessed 22 Sep. 2023].
(91) UNEP. (2021). Seven things you should know about household air pollution. Available at: www.unep.org/news-and-stories/story/seven-things-you-should-know-about-household-air-pollution [Accessed 22 Sep. 2023].
(92) OEP. (n.d) Available at: www.theoep.org.uk/office-environmental-protection [Accessed 22 Sep. 2023].
(93) OEP Press Office. (2023). E-mail to Author, 23 August.
(94) OEP. (2023). Commissioned research to inform OEP's air quality strategy consultation response. Available at: www.theoep.org.uk/report/commissioned-research-inform-oeps-air-quality-strategy-consultation-response [Accessed 22 Sep. 2023].
(95) Ricardo. (2023). Ricardo air quality and economic expertise helps UK Government assess damage caused by air pollution. Available at: www.ricardo.com/en/news-and-insights/insights/ricardo-air-quality-and-economic-expertise-helps-uk-government-assess-damage-caused-by-air-pollution [Accessed 22 Sep. 2023].
(96) https://ieqguidelines.org/table.html [Accessed 24 Sep 23].
(97) IAQM. (2021). 'Indoor Air Quality Guidance: Assessment, Monitoring, Modelling and Mitigation' (version 1.0). Institute of Air Quality Management, London.
(98) www.iaqm.co.uk [Accessed 24 Sep 23].
(99) IAQM. (2020). 'Indoor Air Quality Assessment Levels.' (version 1.0) [pdf]. Available at: https://iaqm.co.uk/wp-content/uploads/2013/02/IAQM_indoor_air_quality_assessment_levels.pdf [Accessed 22 Sep. 2023].
(100) BRE. (n.d). Indoor and outdoor air pollution – how can Home Quality Mark reduce the number of deaths? Available at: https://bregroup.com/buzz/indoor-and-outdoor-air-pollution-how-can-home-quality-mark-reduce-the-number-of-deaths/ [Accessed 22 Sep. 2023].
(101) World Health Organization. Regional Office for Europe. (2022). Measures to reduce risks for children's health from combined exposure to multiple chemicals in indoor air in public settings for children with a focus on schools, kindergartens and day-care centres: supplementary publication to the screening tool for assessment of health risks from combined exposure to multiple chemicals in indoor air in public settings for children. Available at: https://iris.who.int/handle/10665/354225. [Accessed 22 Sep. 2023].
(102) World Health Organization. Regional Office for Europe. (2021). Chemical pollution of indoor air and its risk for children's health: educational course: supplementary publication to the screening tool for assessment of health risks from combined exposure to multiple chemicals in indoor air in public settings for children. Available at: https://iris.who.int/handle/10665/341984. [Accessed 22 Sep. 2023].
(103) WHO. (2010). WHO guidelines for indoor air quality: Selected pollutants. Available at: www.who.int/publications/i/item/9789289002134 [Accessed 22 Sep. 2023].
(104) WHO. (2018). WHO housing and health guidelines. Available at: www.who.int/publications/i/item/9789241550376 [Accessed 22 Sep. 2023].
(105) Heseltine E & Rosen J, eds. (2009). WHO Guidelines for Indoor Air Quality: Dampness and Mould. [pdf] Executive Summary, pp. xii–xvi. Available at: www.euro.who.int/document/E92645.pdf [Accessed 22 Sep. 2023].
(106) WHO. (2021). What are the WHO Air quality guidelines?: Improving health by reducing air pollution. Available at: www.who.int/news-room/feature-stories/detail/what-are-the-who-air-quality-guidelines [Accessed 22 Sep. 2023].
(107) Saffell, J. and Nehr, S. (2023). 'Improving indoor air quality through standardization.' *Standards.* 3(3), 240–267. Available at: https://doi.org/10.3390/standards3030019 [Accessed 22 Sep. 2023].
(108) CEN. (n.d). Available at: www.cencenelec.eu/about-cen/ [Accessed 22 Sep. 2023].
(109) CIB. (n.d). Available at: https://cibworld.org/ [Accessed 22 Sep. 2023].

(110) EPB CENTER. (2019). Interpretation of the requirements in EN 16798-1 — Indoor environmental input parameters for design and assessment of energy performance of buildings addressing indoor air quality, thermal environment, lighting and acoustics (Module M1-6*)* Available at: https://epb.center/support/documents/centr-16798-2/ [Accessed 22 Sep. 2023].

(111) Olesen, B. W. (2017). Indoor environmental input parameters for the design and assessment of energy performance of buildings. Available at: www.rehva.eu/rehva-journal/chapter/indoor-environmental-input-parameters-for-the-design-and-assessment-of-energy-performance-of-building [Accessed 22 Sep. 2023].

(112) CEN. (n.d) Available at: www.cencenelec.eu/about-cen/ [Accessed 22 Sep. 2023].

(113) iTeh Standards. Available at: https://standards.iteh.ai/catalog/tc/cen/8d2afa77-9367-41b4-862d-fbd106461c00/cen-tc-264 [Accessed 23 Sep. 2023].

(114) CENELEC. (2015) *CEN/CENELEC European Guide on Standards and Regulation – Better Regulation through the use of voluntary standards – Guidance for policy makers.* [pdf] Edition 1. Available at: www.cencenelec.eu/media/Guides/CEN-CLC/cenclcguide30.pdf [Accessed 23 Sep. 2023].

(115) European Commission. (n.d) *Level(s) European framework for sustainable buildings.* Available at: https://environment.ec.europa.eu/topics/circular-economy/levels_en [Accessed 23 Sep. 2023].

(116) Dodd, N. et al (2017) 'Level(s) – A common EU framework of core sustainability indicators for office and residential buildings: Parts 1 and 2: Introduction to Level(s) and how it works (Beta v1.0)' Publications Office of the European Union, Luxembourg. Available at: https://publications.jrc.ec.europa.eu/repository/handle/JRC109285 [Accessed 23 Sep. 2023].

(117) Dodd, N., Donatello, S. and Cordella, M. (2021) *User manual: introductory briefing, instructions and guidance (Publication version 1.1)* [pdf]. Available at: https://susproc.jrc.ec.europa.eu/product-bureau/sites/default/files/2021-01/UM3_Indicator_2.4_v1.1_18pp.pdf [Accessed 23 Sep. 2023].

(118) ANEC. (n.d). Available at: www.anec.eu/ [Accessed 23 Sep. 2023].

(119) ECOS. (n.d). Available at: https://ecostandard.org/ [Accessed 23 Sep. 2023].

(120) Umwelt Bundesamt. (2021) International Conference: Limiting Health Impacts of Construction Products regarding VOCs. Virtual Conference, April 20–21.

(121) Antonia, R. and Dirk, J. (2021) 'Harmonised VOC Emission Classes for Construction Products.' [pdf]. *German Environmental Agency.* Available at: www.umweltbundesamt.de/sites/default/files/medien/479/publikationen/doku_05-2021_voc_emission_classes_0.pdf [Accessed 23 Sep. 2023]

(122) ANSES. (2011). *Proposition de valeurs guides de qualité d'air intérieur Méthode d'élaboration de valeurs guides de qualité d'air intérieur.* [pdf]. Available at: www.anses.fr/fr/system/files/AIR2010SA0307Ra.pdf [Accessed 23 Sep. 2023].

(123) ANSES's List of Indoor Air Quality Guideline Values. [pdf] (2018). Available at: www.anses.fr/fr/system/files/Tableau_VGAI_Juillet2018EN.pdf [Accessed 23 Sep. 2023].

(124) Langer, S. (2017). Perceived indoor air quality and its relationship to air pollutants in French dwellings. *Indoor Air*, 27(6), pp.1168–1176. https://doi.org/10.1111/ina.12393 [Accessed 23 Sep. 2023].

(125) Kirchner, S. (2009) 'Indoor air quality in French dwellings.*' International Energy Agency* [pdf] Available at: www.aivc.org/sites/default/files/members_area/medias/pdf/CR/CR12%20IAQ%20French%20Survey.pdf [Accessed 23 Sep. 2023].

(126) Daumling, C. (2013). Hazardous Chemicals in Products - The need for enhanced EU regulations. Brussels, October 29. Available at: www.anec.eu/attachments/UBA_Indoor%20emissions_ANEC-ASI%20CC%20conference%202013.pdf [Accessed 23 Sep. 2023]

(127) Scutaru, A. M. and Witterseh, T. (2020). 'Risk mitigation for indoor air quality using the example of construction products – efforts towards a harmonization of the health-related evaluation in the EU.' *International Journal of Hygiene and Environmental Health*, 229, 113588. Available at: https://doi.org/10.1016/j.ijheh.2020.113588 [Accessed 23 Sep. 2023].

(128) De Brouwere, K. et al. (2020). 'Establishing target and intervention guidance values for indoor air in dwellings and publicly accessible buildings: The Flemish approach'. *International Journal*

of Hygiene and Environmental Health, 230, p.113579. Available at: https://doi.org/10.1016/j.ijheh.2020.113579 [Accessed 23 Sep. 2023].

(129) Official Journal of the European Union. (2021). *Regulations: Commission Delegated Regulation (EU) 2021/21 39.* [pdf] Available at: https://eur-lex.europa.eu/legal-content/EN/TXT/PDF/?uri=CELEX:32021R2139 [Accessed 23 Sep. 2023].

(130) HPI. (n.d). *Technical Documents.* Available at: https://homeperformanceindex.ie/technical-manual-download/ [Accessed 23 Sep. 2023].

(131) MKN Property Group. (n.d). Available at: www.mknpropertygroup.com/ [Accessed 23 Sep. 2023]

(132) Castlethorn. (n.d). Available at: www.castlethorn.ie/helpful-guides [Accessed 23 Sep. 2023].

(133) The Ella Roberta Foundation https://ellaroberta.org/about-ella [Accessed 24 Sep. 23].

(134) Shared Regulatory Services. (2022) *Bridgend 2022 Air Quality Progress Report: In fulfilment of Part IV of the Environment Act 1995.* [pdf] Available at: https://democratic.bridgend.gov.uk/documents/s28416/Bridgend%20APR%202022%20final%20DRAFT%20JB%20comms%20v1.2.pdf [Accessed 23 Sep. 2023].

(135) Clean Air Programme. (n.d) Available at: www.ukcleanair.org/projects/?wave=wave-1 [Accessed 23 Sep. 2023].

(136) Clean Air Programme. (2021). 'TRANSITION Clean Air Network – Optimising air quality and health benefits associated with a low-emission transport and mobility revolution in the UK' Available at: www.ukcleanair.org/projects/optimising-air-quality-and-health-benefits-associated-with-a-low-emission-transport-and-mobility-revolution-in-the-uk/ [Accessed 23 Sep. 2023].

(137) Tang, R. and Pfrang, C. (2023). 'Indoor particulate matter (PM) from cooking in UK students' studio flats and associated intervention strategies: evaluation of cooking methods, PM concentrations and personal exposures using low-cost sensors.' *Environmental Science: Atmospheres,* 3, 537–551. Available at: https://research.birmingham.ac.uk/en/publications/indoor-particulate-matter-pm-from-cooking-in-uk-students-studio-f [Accessed 23 Sep. 2023].

(138) Pfrang, C. (2021). Air Pollution Solutions for Vulnerable Groups (CleanAir4V). Available at: www.ukcleanair.org/projects/air-pollution-solutions-for-vulnerable-groups-cleanair4v/ [Accessed 23 Sep. 2023].

(139) Global Action Plan. (n.d). Clean Air Hub. Available at: www.cleanairhub.org.uk/home [Accessed 23 Sep. 2023].

(140) Hudson, B. (2022). Engaging the Public: Indoor & Outdoor Air Quality Research. Available at: www.ukcleanair.org/projects/engaging-the-public-indoor-outdoor-air-quality-research/ [Accessed 23 Sep. 2023]

(141) BIOAIRNET. (n.d). Indoor/Outdoor Bioaerosols Interface and Relationships Network. Available at: www.bioairnet.co.uk [Accessed 23 Sep. 2023].

(142) Coulon, F. (2021). BioAirNet – Indoor/ Outdoor Bioaerosols Interface and Relationships Network. Available at: www.ukcleanair.org/projects/indoor-outdoor-bioaerosols-interface-and-relationships-network-bioairnet/ [Accessed 23 Sep. 2023].

(143) HEICCAM. (n.d). Health and Equity impacts of Climate Change Mitigation measures on indoor and outdoor air pollution exposure. Available at: heiccam.org [Accessed 23 Sep. 2023].

(144) HEICAM Network. (n.d). Written evidence by the HEICCAM Network. Available at: https://committees.parliament.uk/writtenevidence/121551/pdf/ [Accessed 23 Sep. 2023].

(145) Roberts, K. (n.d) 'The need for a national clean air strategy for schools.' Education Business. Available at: https://www.educationbusinessuk.net/features/need-national-clean-air-strategy-schools [Accessed 23 Sep. 2023].

(146) TAPAS. (n.d). Available at: https://tapasnetwork.co.uk/ [Accessed 23 Sep. 2023].

(147) Immaterial. (n.d). Available at: www.immaterial.com [Accessed 23 Sep. 2023].

(148) Sturgeon, T. (2022). Platform technology for the removal of critically underserved pollutants in homes. Available at: www.ukcleanair.org/projects/platform-technology-for-the-removal-of-critically-under-served-pollutants-in-homes/ [Accessed 23 Sep. 2023].

(149) National Engineering Policy Centre. (n.d). Ventilation matters – why clean air is vital to health. Available at: https://explainers.raeng.org.uk/ventilation-matters [Accessed 23 Sep. 2023].

(150) Breathing City. (n.d). Future Urban Ventilation Network. Available at: www.breathingcity.org [Accessed 23 Sep. 2023].

(151) Thornton, C. (2022). Consortia: RESPIRE – Relating Environment-use Scenarios in Pregnancy/Infanthood and Resulting airborne material Exposures to child health outcomes. Available at: www.ukcleanair.org/projects/respire-relating-environment-use-scenarios-in-pregnancy-infanthood-and-resulting-airborne-material-exposures-to-child-health-outcomes/ [Accessed 23 Sep. 2023].

(152) Moghaddam, R. S. (2022). CodiKoat: Harnessing Nanoparticle array technology for the removal of domestic atmospheric pollutants. Available at: www.ukcleanair.org/projects/codikoat-harnessing-nanoparticle-array-technology-for-the-removal-of-domestic-atmospheric-pollutants/ [Accessed 23 Sep. 2023].

(153) CODIKOAT. (n.d). Available at: www.codikoat.com [Accessed 23 Sep. 2023].

(154) ARBNCO. (n.d). Available at: www.arbnco.com [Accessed 23 Sep. 2023].

(155) Stewart, A. (2022). Measure, Inform, Nudge: an integrated, human-centric air quality measurement & visualisation system. Available at: www.ukcleanair.org/projects/measure-inform-nudge-an-integrated-human-centric-air-quality-measurement-visualisation-system/ [Accessed 23 Sep. 2023].

(156) Lynn, C. (2022). FamilyAIR: Personalised, Actionable Air Quality Monitoring for the Entire Family. Available at: www.ukcleanair.org/projects/familyair-personalised-actionable-air-quality-monitoring-for-the-entire-family/ [Accessed 23 Sep. 2023].

(157) Higgs, V. (2022). Smarter Home Indoor Air Quality Monitoring System. Available at: www.ukcleanair.org/projects/smarter-home-indoor-air-quality-monitoring-system/ [Accessed 23 Sep. 2023]

(158) McFiggans, G. (2021). Consortia: A new pollutant hazard platform. Available at: www.ukcleanair.org/projects/a-new-pollutant-hazard-platform/ [Accessed 23 Sep. 2023].

(159) Carslaw, N. (2021). Consortia: Ingenious – Understanding the sources, transformations and fates of indoor air pollutants. Available at: www.ukcleanair.org/projects/air-quality-and-urban-homes/ [Accessed 23 Sep. 2023].

(160) Imperial College London. (n.d). WellHome – West London Healthy Home and Environment Study. Available at: www.imperial.ac.uk/school-public-health/environmental-research-group/the-wellhome-study/news/ [Accessed 23 Sep. 2023].

(161) Declaration of interest: the author has attended a "stakeholder" meeting of this project.

(162) Bunn, S. and Duffield G. Indoor Air Quality 26.9.23 Post Brief 54 UK Parliament https://post.parliament.uk/research-briefings/post-pb-0054/

(163) TCPA Healthy Homes Campaign. (2023). Available at: www.hometodiefor.co.uk/ [Accessed 3 Oct. 2023].

(164) Levelling up and Regeneration Bill. (2023). Available at: https://bills.parliament.uk/publications/52700/documents/3949 [Accessed 3 Oct. 2023].

5 Damp, mould, building materials and retrofit

There is a growing recognition of problems of mould in buildings. This is often referred to as a problem caused by occupants, who fail to ventilate their houses properly, or create too much moisture. It is easy for experts to blame the occupants so they avoid recognising the real problem, that modern building methods and materials have led to much greater problems with mould. Mould is a health hazard caused by a complex combination of poor ventilation, high relative humidity and inappropriate building materials, which is why it is included in this book. It is essential to understand the ways in which building materials contribute to mould.

Having become aware of damp and mould, a young architecture student in Edinburgh, in the late 1960s, dissatisfied with their course, began to develop what became known as community architecture. Initial work began in Craigmillar, a large housing estate where there were many council houses in poor condition. The Craigmillar Festival Society ran a very successful musical and historical festival to raise the spirits of local people, but also did remarkable work helping people living in poverty and bad housing conditions. Craigmillar continued to exhibit serious problems of dereliction for decades despite various renovation proposals. Even today there is talk of a new masterplan to try and regenerate the place. Plans are still being considered to regenerate Craigmillar 50 years later! **(1)**, **(2)**

Recent media reports show that there are serious problems of mould and damp such as the case of Pauline Jones in Craigmillar.

> Scots mum dies just weeks after daughter complained about horrific mould in flat
>
> Pauline Jones recently passed away aged 57 from kidney failure, leaving behind her doting daughter Andrea Jones. **(3)**

Fifty-five years after finding damp and mould in Edinburgh, the problem is still with us.

Community technical aid

The term community architecture was taken over by the Royal Institute of British Architects with the help of Prince (now King) Charles, in an effort to thwart radical grass roots activity supported by young radical architects, though the Royal Institute of British Architects (RIBA) version of community architecture was largely talk. **(4)**

So, in the mid-70s what became known as the Community Technical Aid Movement was set up as an alternative to the ideas of Charles and the RIBA. At its peak, the Association of Community Technical Aid Centres had well over a hundred-member organisations, who were

DOI: 10.1201/9781003129226-5

providing architectural and technical advice to community groups, particularly around issues of mould and bad housing. **(5)**

There were several alternative architecture and technical aid agencies set up in Glasgow, the earliest being "Assist," and there were appalling conditions of mould in private and council housing estates where the practice, which continues to this day, of blaming the tenants and occupants for causing damp and condensation was established. There were many campaigns in Scotland at this time some of which have been documented by Valerie Wright, including the story of the "Dampness Monster. **(6)**, **(7)**, **(8)**

In the 1970s in London, another of the early technical aid groups, "Support" worked with London law centres to take landlords to court, when tenants were found to have severe mould and damp problems.

Attention then turned to tower blocks and over the next fifteen years many tenants' campaign groups succeeded in getting many blocks demolished. **(9)**

Despite several books on the subject it's hard to get an accurate figure of how many high-rise and "deck access" blocks were demolished, maybe 1500 or more, including Ronan Point where there was a gas explosion in 1968. **(10)**

Most of the flats in these buildings suffered from damp and mould through water ingress and condensation, due to minimal levels of insulation and inadequate heating, problems which continue to today.

Debunking the "blame the tenants" approach

A groundbreaking study in 1985 by Markus and Nelson at Strathclyde University about the Darnley Housing Estate in Glasgow provided a comprehensive analysis of the factors creating condensation and mould. The council landlord, as usual, was happy to blame the tenants and tried to ban the use of propane gas heaters (which can create high levels of moisture) but the building science showed clearly that exposure to wind-chill, with minimal fabric insulation, was the main cause of the damp problems. **(11)**

Despite this study being carried out in the early 1980s the conclusions remain much the same today. The conditions in houses that had been "improved" were worse than the ones that were not improved and an electric heating system was found to be unsuitable. The study points out that condensation problems were largely due to the poorly insulated construction and what they called climatic severity. The study also found considerable difference in the heating used by occupants as some couldn't afford heating at all. The occupants kept diaries so there was a great deal of information on "moisture producing activities." This made it possible to discount occupant behaviour as a significant variable.

Much social housing in Scotland was heated by electricity, beginning the spiral into fuel poverty so familiar today. Once again pressure to introduce expensive electric heating under the name of "decarbonisation" has returned, as well as blaming the tenants.

Work on other tower blocks, such as at Roystonhill in Glasgow, where external cladding had been installed, featured in a BBC TV documentary, "Cause for Concern" presented by Margo McDonald in 1987, displaying high levels of mould and structural defects. Similar problems were investigated by the BBC in East Kilbride nearly 40 years later where tenants were complaining that mould had resulted from retrofit insulation measures. **(12)**

It took the gas explosion at the Ronan Point flats in east London to accelerate the demolition of many multi-storey housing blocks, but many remain, to this day, and the unsatisfactory nature of these blocks and attempts to retrofit them led to the Grenfell fire, as explained by Inside Housing journalist Peter Apps. **(13)**

Poor construction, inadequate insulation, heating and ventilation is also a problem affecting many low-rise buildings as in East Kilbride, not only housing owned by Local Authorities and Housing Associations, but private landlorded property and even private owner-occupied housing. High levels of mould and other indoor air quality problems can be found in the houses of well-to-do people and also in many other non-domestic buildings such as schools and workplaces. One of the main solutions offered for damp problems today is to internally line walls with PIR insulation, the same material that burned on Grenfell tower. This introduces higher level emissions from flame retardants, isocyanates, glues and increases fire risks into the inside of the houses of vulnerable people.

During March and April 2023, the author appeared on national BBC TV news and several local radio stations, in programmes about residents suffering from damp, as people bombarded the media with stories of bad conditions which had not been attended to by landlords. There is no space here to document all these cases but here is one example:

> A baby was hospitalised with breathing difficulties "because of damp and mould" at a family's Laindon home in Essex. Francesca Clark, 31 and her partner Andy Coleson, 38, claim their Russell Close flat is "infested with mould". The Clarion Housing tenants say their seven-month-old son, Teddy, was hospitalised as a result of the conditions. The youngster was taken to Basildon Hospital on Thursday, January 6 (2022where he was kept under observation for 24 hours. The couple claim doctors told them his illness was caused from living in damp and mouldy conditions. A Clarion (Housing Association) spokesman said: "We take every case of condensation, damp and mould very seriously and we are committed to a permanent resolution to the issue in Ms Ward's home." As soon as we learned of Ms' Ward's son's health issues we arranged for an immediate move. **(14)**

This family was moved to a "cramped and unsuitable" hotel room. It has been suggested that there is a global epidemic of mould, which can be found in the UK, temperate, tropical and sub-tropical countries around the world. Anywhere where there are high levels of relative humidity, such as Cape Town in South Africa, which has an army of white vans going around the city tackling mould! **(15), (16)**

Treating mould has been with us for thousands of years

Mould spores are everywhere and get the opportunity to breed thanks to unsatisfactory building materials and methods and conditions which create condensation. While landlords continue to blame occupants for engaging in "anti-social activities" such as breathing, washing and cooking, the real villain is to be found in the building fabric. Following the tragic death of Awaab Ishak in Rochdale in 2020 the "Levelling-Up" Minister Michael Gove has changed legislation intended to protect tenants and instructed civil servants to update mould and damp guidance but the drafts and minutes of the "experts" assembled to address this issue reveal how little is understood about mould. The problem has been around for 3,000 years! **(17)**

Advice from above!

In the Old Testament of the Judaeo-Christian bible, around 1000 years before Christ, the problem of mould is set out clearly, though with some rather bizarre suggestions for remedies. In Chapter 14 of the book of Leviticus, Moses was advised by 'God' about the problems of "plague and leprosy" in houses. The use of the words plague and leprosy may well be due to mistranslation, as the chapter includes clear references to building defects and there are various different translation versions of verses 33 to 53. However, it is clear in Leviticus that priests were given technical advice on how to get rid of "spots" and to re-plaster the house (with earth and probably lime). Houses were also to be purified using "cedar wood, scarlet and hyssop" and also the blood of a living sparrow, which was allowed to fly away once the house had been purified, though some versions involve dead sparrows! Cedar wood is used to this day in biocidal treatments.

Mould or mildew was apparently placed upon a house by "God" as a punishment for sinners, in some religious interpretations, and this might explain the basis for the incompetence of landlords and housing managers even to this day. The reader can decide who are the sinners, the tenants, landlords or builders! **(18)**

Is creating bad air quality criminal?

As coroners have ruled that children, including Awaab and Ella Kissi Debra in London, have died as a result of mould growth and air pollution from traffic, this has led to greater attention on the problems of the health impacts of air pollution. **(19), (20)**

So far, no criminal prosecutions have followed, though establishing negligence cases against housing professionals, architects and builders can be difficult. Seven years after the Grenfell disaster, where 72 people died, prosecutions have still not followed, though the police have been involved in investigating various companies through Operation Northleigh. **(21)**

When the tragic case of Awaab Ishak came to light, the authorities quickly realised that current practices and guidance were unsatisfactory. The "Housing Ombudsman" was called to give evidence at the coroner's inquest but the coroner did not call anyone with expertise in building and the causes of mould. Evidence of the health effects of mould were provided by medical expert Professor Malcolm Richardson from Manchester University. A more recent death of occupants in Oldham and in Liverpool has led to further inquests and police have been asked to investigate. **(22), (23)**

Blame the tenants and occupant behaviour, the academic answer

A literature search on occupant behaviour in relation to fuel poverty and indoor air quality, damp and mould will uncover hundreds of academic papers devoted to the topic of occupant behaviour. Blaming the occupants is not only a result of ignorant managers and professionals trying to deflect blame, but is backed up by a massive but questionable body of academic research. Much of this ignores the environmental and fabric causes of poorly heated houses, full of damp and mould, but puts the focus firmly on occupant behaviour and the need to "educate" people. If only people were persuaded to turn down their thermostat a bit then all would be well!! The landlord in Rochdale was severely criticised by the Coroner for blaming the parents of Awaab, and failing to take any action to address the problem even though they were aware of it. It has also been suggested that this is a result of both class and racial prejudice. **(24), (25)**

Very often academic studies appear to be supportive of people in fuel poverty, for instance, suggesting that they need training and support so that they know how to use new heating systems, following retrofit programmes. But these suggestions frequently adopt a paternalistic and patronising approach to occupants, where the retrofit measures are based on middleclass professional attitudes, rather than a genuine understanding of the needs of poorer people. So extensive is this literature that only a few examples (i.e., Kearns et al 2019, Cayla et al 2011, Blight et al 2013, Monahan et al 2012, Yohannis 2012) are cited here to indicate this. There are many, many, more. **(26), (27), (28), (29), (30)**

Much of the literature maintains a patronising theme suggesting that occupant behaviour has to be changed and that tenants are difficult to educate as suggested by Palmer et al 2028!

> …… .tenants have a significant role in avoiding or sometimes causing condensation problems, recognising that occupant behaviour often needs to change after retrofit work – especially relating to ventilation. Some interviewees said that it is **challenging to educate tenants** about changes they should make when their homes have been improved. **(31**)

A much smaller group of academics has tried to see things from the tenant's perspective fortunately. The rather unfortunately (in this context) named Ronald Mould and Keith Baker provide a number of alternative insights into the problem, based on a more sympathetic approach to occupants including a brilliant critique of energy performance certificates and bad retrofit policies, arguing in favour of a more person centric approach. **(32)**, **(33)**, **(34)**

There is no doubt that occupant behaviour in houses will vary between age groups, classes, cultures, work patterns, income levels and many other factors but houses should be built or renovated robustly to cope with this, rather than people being expected to alter their behaviour to suit buildings and energy policies. An interesting case in Scotland, with a luxury high specification new build housing built in 2022, intended to be ultra-low energy, was found to have serious mould growth well before any occupants had moved in. The condensation risk analysis provided to the architects claimed that there was no risk of mould growth as there would not be any cold surfaces to attract mould, and yet the mould was already well established in the empty apartments, and was treated through fungicidal spraying. Full details cannot be given due to legal reasons, as is often the case where these matters are settled out of court or by insurers. Lessons are not learned from such problems due to non-disclosure agreements. Building fabric that can exacerbate mould should not be blamed on the occupants as it is frequently established before people move in and yet this is becoming a widespread problem.

The death of Awaab Ishak

Awaab died on the 21st of December 2020 at the Royal Oldham Hospital. He lived in a flat in the unfortunately named **Illminster** block, on the Freehold estate in Rochdale. The Inquest opened 30 March 2022 and concluded on 4 November 2022. The medical cause of death was given as Acute Airway Oedema with Severe Granulomatous Tracheobronchitis, but the coroner also recorded that this was the result of environmental mould exposure. The coroner noted that too much emphasis was placed on the cause of the problems being due to parent's lifestyle and the landlord did not identify the lack of an adequate ventilation system as a factor in the presence of the mould. Due to the lack of remedial action by the landlord, Awaab continued to be exposed to harm.

The coroner sent a Regulation 28 Report to the English Government requiring them to respond within 56 days. A Reg 28 report is meant to alert the Government to the need to prevent further deaths. **(35)**

Unfortunately, the coroner's inquest did not properly investigate the building fabric problems, relying largely on medical evidence from Professor Malcolm Richardson. While Richardson is an eminent medical expert on mycology with over 450 publications **(36):**

> In his view, it (mould) had been present for some considerable time, although he could not date it. He explained to the court the importance of locating and understanding the source of the damp or water. For example, in the bedroom cupboard he was of the view that the pattern was likely due to water ingress from **somewhere** but told the court **it was not his role to locate the source. (37)**

Sadly Richardson, and the Housing Ombudsman, who also gave evidence, were not qualified to understand the cause of the mould. In Richardson's extensive medical publications, the link between mould and respiratory problems like COPD is well established, but it was not possible to find a single paper in his work where the source of the mould had been investigated. From his evidence, Richardson suggested that the mould was due to water ingress rather the more likely cause of condensation. While the science of mycolog, in terms of health effects is extensively covered in medical literature, there seems to have been insufficient research into the causes of mould and building materials involved. On a visit to Illminster house it was possible to see that some kind of proprietary polymer cement render external wall insulation (EWI) had been applied to the external walls on all of the blocks in the Freehold estate a few years before. Rochdale Boroughwide housing was asked for details of this work but did not reply, despite several reminders. It was possible to see that this EWI work was already starting to deteriorate in places and there was severe damp staining around downpipes and other defects.

Sealing up the outside of the building with such an EWI system has been shown to cause extensive mould and condensation internally in many other cases, but due to the unwillingness of the landlord to co-operate, further detailed investigation was not possible. From a visual inspection, it was unclear whether any improvement in ventilation had been installed since the coroner's report, even though press reports suggested that this was under way.

The coroner's office also refused to reply to an enquiry about why experts on the causes of damp and mould had not been called to the Inquest. While the decision of the coroner on this case was groundbreaking, it is unfortunate that the court did not seek evidence from experts on damp, mould and building defects. It is possible that the coroner assumed that the Housing Ombudsman who gave evidence was an expert on mould, but it is clear from the Housing Ombudsman's report "Spotlight on damp and mould – Its not lifestyle" (Housing Ombudsman Service Liverpool October 2021) that the Ombudsman team that wrote this report had only a limited knowledge of the subject. Ombudsman chief, Richard Blakeway, does not appear to be an expert on mould and dampness, having been a housing and investment administrator and an election observer in Somaliland according to his CV. The Ombudsman's report fails to address the impact of retrofitting and different insulation methods. The Housing Ombudsman also refused to respond to several email enquiries but in a Freedom of Information Response confirmed that they did not have any technical expertise:

> The members of the Quality, Engagement and Development Team are not required to have technical knowledge of a specific subject, such as damp and mould. The team members are required to have knowledge of housing law and policy. They are required to hold experience into quantitative and qualitative research methods, and have experience in planning and delivering research investigations. They also required to have an understanding of the wider social, political and economic environment and their impact on the issues, implications and challenges facing public sector organisations. **(38)**

They also stated that there are no plans to revisit the published content of their Spotlight report even though they did publish a second report in which they say

> The volume of casework and findings have increased significantly. The number of findings we made about the handling of damp, mould and leaks increased from 195 in 2020-21 to 456 in 2021-22, a 134% increase. The report highlighted extensive maladministration among social housing landlords, but did not refer to the causes of mould and damp or any work that may be being carried out to remedy the problems. **(39)**

In response to the FOI they also admitted that they didn't directly consult any experts on mould and damp, not revealing whether they had received any technical information as a result of a call for evidence in relation to their "Spotlight" work. The Ombudsman have been doing good work to investigate problems of damp and mould and they publish judgements against social landlords every week (see footnote) but they are not addressing the technical causes of mould.

> In addition to reviewing our casebook for the last two financial years, we also conducted a call for evidence that ran for seven weeks during April to June 2021, asking for assistance from both the public and sector professionals to inform our understanding. We held discussions with landlords and with several representative bodies, including the National Housing Federation, the G15 organisation and the Northern Housing Consortium. We also held discussions with our Resident Panel and the Tenant Participation Advisory Service. **(40)**

> In 2021–22, 13 of the 31 severe maladministration decisions we made were about the handling of damp and mould reports. The landlords involved ranged in size from just over 1,000 homes to almost to 110,000. Of these, four were medium sized landlords of between 1,000 and 10,000 homes and nine are responsible for more than 10,000 homes. Six were local authority landlords and seven were housing associations. This indicates the issues are widespread across the sector, regardless of the size or type of landlord.
>
> The volume of casework and findings have increased significantly. The number of findings we made about the handling of damp, mould and leaks increased from 195 in 2020–21 to 456 in 2021–22, a 134% increase.
>
> The rate at which we upheld those findings increased from 37% to 45%. We had 1,993 enquiries and complaints about damp, mould and leaks in 2020–21 – that figure increased last year to 3,530, a 77% increase and as of December 2022, we had already received 3,969 enquiries and complaints for 2022–23.

One year on follow-up report: Housing Ombudsman Service, 2 February 2023

Another official body, the Housing Regulator, also fails to investigate the cause of the problems that are leading to the massive amount of mould and damp problems. The Housing Regulator issued a 3-page "report" that states rather limply that

> "most social landlords understand the extent of damp and mould in their tenants' homes and take action to tackle it but could strengthen their approach further. We will expect all registered providers to make improvements to how they protect tenants from the potential harm that damp and mould can cause.

Data released by the Housing Regulator vastly underestimates the extent of damp and mould in the UK, and only reported cases identified by social landlords, ignoring mould in private rented and owner-occupied property. **(41)**

Failing to understand the problem

One of the dangers of having only a partial understanding of such problems is that the real lessons are not learned. This was clearly shown from the Lakanal House Coroner's report about the fire there in 2009 where six people died which failed to fully understand the cause of the fire and as Peter Apps claims could have prevented the Grenfell disaster. The Lakanal Coroner did call on the

Government to review fire safety, but this was never properly followed up, and as Apps explains, did not fully explain the nature of the problems that led to the deaths of those six people. **(42)**

As the coroner's report about Awaab's death report does not address the real cause of the mould growth, we do not have the full story about the issue. The Manchester Evening news reported widespread mould problems in other flats on the estate making clear that the problem has not been adequately addressed. **(43)**

Why do housing "experts" not understand the problem – training?

While a coroner has ruled that children have died as a result of mould growth and air pollution (Awaab and Ella) housing managers, surveyors and landlords do not appear to be held accountable for their ignorance about mould and poor air quality.

From experience teaching housing managers and others in the 80s about building technology, professionals at least had some awareness of these problems but many of the issues referred to in this book no longer form part of the curriculum for surveyors, housing managers and environmental health officers. Even building control officers who might be expected to police bad practice seem happy to approve toxic, flammable and mould-inducing materials in buildings.

The training of housing managers approved by their professional body, the Chartered Institute of Housing (CIH), used to include modules on building technology but this no longer is the case. Environmental health officers also receive very little training on building materials and related issues like damp and mould. This means that when confronted with problems such as mould growth, housing managers rely on amateurish advice about cleaning off mould with bleach and have little understanding of what has caused the problems. When offered some simple training events that could demystify technical issues the CIH responded by saying they would not get enough takers to pay for such courses as housing managers would not want to address technical issues[1].

In response to an email to the Chartered Institute of Housing (CIH), a senior official admitted,

> there has been a move away from building and technical issues in housing management qualifications, particularly with the demise of more in-depth housing specific graduate and post-graduate programmes. **(44)**

Advice about damp and mould provided by the CIH is very poor with no attempt to understand or address the cases of condensation or mould, while appearing to be sympathetic to tenants.

> Treat the problem – the response will vary depending on severity, but always focus on resolving the root cause. This can include giving residents sensitive advice on how to minimise the build-up of condensation, without blaming the issue on lifestyle factors. **(45)**, **(46)**

A UK Government announcement on February 26, 2023 revealed the surprising information that 25,000 managers in the housing sector are not professionally qualified.

> Professional qualifications to be made mandatory for social housing managers, ensuring residents receive a high-quality service and rapidly professionalise the sector. Social housing managers must gain professional qualifications under new rules to protect residents and raise standards in the sector, Housing Secretary Michael Gove announced today. Around 25,000 managers across the sector will now be required to have an appropriate level housing management qualification recgulated by OfQual equivalent to a Level 4 or

5 Certificate or Diploma in Housing, or a foundation degree from the Chartered Institute of Housing. **(47)**

Lack of guidance from Government about indoor air quality in relation to mould

The DEFRA report on indoor air quality (discussed in Chapter 4), for instance, may contribute to, rather than reduce, the confusion around the causes and effects of mould with 16 references to the subject scattered throughout their 142 pages. DEFRA does cite Johansson who has also published an important study of the effect of thin renders on external insulation causing mould internally.

> The key factor influencing the respirable burden of mould-forming fungi and bacteria is the presence of excess moisture on surfaces and in the air (WHO, 2009). This in turn is influenced by the type of materials used in the home that may support biological growth to different extents. Dampness not only initiates the degradation of materials in the home, it allows the growth of microbes which then pollute the air with their spores, allergens, volatile compounds and toxins (see 3.1.5). Accordingly, controlling sources of dampness can lead to health improvements, such as has been shown by intervention studies whereby removing sources of dampness has reduced asthma exacerbations in highly symptomatic children while surfaces can generally be kept free of moulds if the relative humidity is kept below 75%. Growth of moulds is influenced by the characteristics of specific materials; critical relative humidity for growth on construction materials varies from 75-80% for wood and wood-based materials, 80-85% for paper on plasterboard and 90-95% for polystyrene and concrete (Johansson et al., 2005). **(48)**

Mould is also attracted by plastic materials such as UPVC as anyone with UPVC windows will be aware. The following statement in the DEFRA IAQ report indicates further misunderstandings of the issues. Condensation and mould is not just a problem of water ingress:

> Managing microbial growth through controlling relative humidity can be best achieved by addressing moisture-related problems in buildings including leaking, flooding, water wicking through porous building materials, unvented sources (bathrooms and laundry rooms) and condensation on cooler surfaces such as windows and thermal bridges in insulation. Many of these issues are addressed by better building codes for new buildings, and can also be mitigated through behavioural and mechanical interventions in existing buildings. **(49)**

While the new Government bill was going through Parliament, an advisory expert group was hastily convened to produce guidance on damp and mould in buildings with plans to issue a guidance document in July 2023. A new Government Department entitled Office for Health Improvement and Disparities published a draft guidance document for consultation during June and July. It is puzzling how this "expert" advisory panel was assembled. For instance, it includes an academic with only two publications (listed from 2014 and 2016), neither of which are related to dampness and mould, another academic with an interest in wellbeing but no published work on damp and mould, with most other members being senior officials with some relevance to the topic, but unlikely to have any technical expertise. There were only three academics with some relevant expertise. Not surprisingly the draft report contained a number of serious errors and clear evidence of a lack of technical knowledge of damp and mould by whoever was drafting it. For instance, a great deal of emphasis was placed on rising damp even though it is relatively uncommon, with penetrating damp as the draft correctly point out, caused by various building defects such as faulty downpipes and gutters and lack of drainage at the foot of walls.

> **Rising damp** … . is not common, usually found in old properties, and is often misdiagnosed. It can be identified through visual inspection. **(50)**

The draft failed to include guidance on non-toxic cleaning materials in the section on removing mould though it did include "*specifying materials that are breathable and easy to clean.*" However, this could cause confusion as the term breathable should be Vapour Permeable, and easy to clean should be a separate category and should not be confused with breathability. The draft did not provide details of vapour-permeable materials and in the end no reference to breathable was included in the final document. Easy to clean was also removed from the final version.

The draft then stated wrongly that "Modern buildings are now normally constructed with breathable materials."

> In older buildings, newer finishes might have been used and contributed towards decreasing vapour permeability and making 'breathing' more difficult. In addition to ensuring adequate ventilation, if appropriate, landlords may wish to use waterproof coatings or render options to maintain vapour permeability in such buildings.

Most modern buildings and retrofit schemes use **vapour-closed** materials, and this is one of the main reasons for the increase in mould. Some synthetic insulation materials such as mineral fibre can be considered as vapour permeable but most commonly used plastic foam insulations are not. Most conventional finishes using gypsum plasters, cement renders and plastic paints are not vapour permeable. This was also deleted from the final draft so there was no discussion of breathability or vapour open materials.

The final version of the guidance on damp and mould was eventually issued by Government on September 7th having been promised in mid-July. **(51)**

It is a lengthy document mainly focussed, quite rightly, on protecting tenants and making clear the obligations on landlords. Most of the legal and policy measures listed such as HHSRS are not new, and it has been possible to force landlords to deal with mould and damp for decades through legal action if funds are available. It's not certain that this document will make much of a difference as landlords, dragging their heels, can ignore the needs of tenants. Many people are worried about becoming homeless and so tend not to complain even when conditions are very poor. The policy does not include any resources to establish local agencies that tenants can contact to get advice about their mould.

The document says that tenants should not be blamed for damp and mould "which are not the result of lifestyle choices," but much later on the document says

> to address condensation involves working with tenants to make small, reasonable adjustments to their **behaviour,** if appropriate, to reduce their damp and mould risk.

This suggests that "blame the tenants" is still lurking in the minds of policymakers.

Explaining the causes of damp and building defects has been slightly improved from earlier drafts. For instance, rising damp, which is rarely the cause of mould, is still in the policy but referred to as "often misdiagnosed."

Retrofit measures are discussed in the policy but without any explanation of unintended consequences, relying on the highly flawed PAS 2030 and 2035. It is assumed that retrofit installers, if they are Trust Mark certified, will avoid dangers of mould. The policy completely fails to address the fact that much of the serious existing mould problems are a result of previous botched and unsatisfactory retrofit schemes and because PAS 2035 fails to address the detailed nature of insulation measures are likely to continue to occur in the future.

Sealing up houses with plastic-based, vapour-closed insulation measures have been one of the biggest causes of mould for many years, and this is not discussed in the document at all. Insulation measures are recommended a few times but without any detailed explanation and landlords are encouraged to consider "improving external and internal wall insulation," without any further detail. This is the most serious flaw in the guidance and renders the whole document almost worthless as the implication is just, business as usual, without a recognition of the damage that can be done. Despite apparently being sympathetic to tenants and setting out bureaucratic measures to implement existing legislation, this guidance will completely fail to safeguard vulnerable people from becoming ill due to mould. The named people involved in advising the Government listed in the report have failed to warn of the dangers of retrofit disasters and have done a great disservice to the thousands of people suffering from mould in England. The new guidance was issued very quietly with no media coverage.[2]

Health problems as a result of mould

There are many thousands of species of mould, and not all are as dangerous to health as some, but mycotoxins are seen as particularly dangerous. Richardson et al say that indoor environments contain about 100–150 species, a small fraction of the more than 100,000 species of described fungi. They state that *Aspergillus fumigatus* has an environmental tenacity and thermo tolerance, and so is found in diverse ecological spaces.

> However, little is known about the micro-environmental and biotic niches in indoor environments that support the growth of Aspergillus. **(52)**

The term toxic mould is commonly used and are referred to as MVOCs in indoor air literature.

> Certain moulds are toxigenic, meaning they can produce toxins (specifically "mycotoxins"). Hazards presented by moulds that may produce mycotoxins should be considered the same as other common moulds which can grow in your house. **(53)**

Some indoor air quality experts are keen to identify different species of mould when carrying out house inspections, but it is safe to say that occupants will experience health problems if any form of mould is found.

Dust mites and damp-related bugs

Constant high humidity can lead to serious problems with dust or mould mites Tyrophagus putrescentiae, and many people suffer from allergies from infestations. Building emissions deniers talk about particulates at great length and dust in houses can be a problem, but it is the mites that really cause some health problems. The tiny mites (which can be visible to the naked eye) feed on mould and they breed very rapidly. Absorbent materials such as mattresses and furnishings, which have high moisture content, will provide a breeding ground but so do certain building materials that attract mould as the mites need mould to survive.

Research in schools in Finland relate a wide range of illnesses to conditions in a moisture damaged school. These include abdominal pain, nausea, diarrhoea, vomiting, respiratory symptoms such as cough, dyspnea, difficulty in breathing, problems with concentration, brain fog, memory difficulties, dizziness, hearing problems and ear infections, balance problems, headache, eye irritation, rash, asthma, chemical sensitivity and even cardiac problems.

> Indoor air toxicity and dampness-related microbiota recovered from the classrooms were associated with multi-organ morbidity of the school occupants.

This paper is interesting as it tackles the issue of causality, citing numerous other studies of the links between damp mould and a wide range of illnesses including neurological and chemical sensitisation and claims that,

> spending a long time in a moisture-damaged building results in morbidity, that is irrefutable. There is now an awareness that fungi can expel tiny liquid droplets that aerodynamically migrate into the airways in conditions of increased relative humidity. **(54)**

There is a tendency to assume that health problems only occur when mould is clearly visible, but in reality, mould can be well established, but out of sight, and health problems can develop before the more obvious black growth can be seen. Norback et al have drawn attention to the problem of invisible mould citing studies showing associations between dampness in the floor construction and airway obstruction. This illustrates that emissions from building dampness that are not related to visible mould growth could influence lung function. They also state that women are more vulnerable to ill health from mould. They show that mould is related to microbial compounds and endotoxins from bacteria and chemicals which can be emitted from damp building materials such as 2-ethyl-1-hexanol. **(55)**, **(56)**

There is a major gap between those who carry out medical research and those that understand buildings and materials. UK literature on the problem of mould and health very rarely identifies the physical problems in buildings, and there is a lack of epidemiological information. While allergic and other reactions to fungi are well known, more work is needed, particularly where these are seen as a source of VOC emissions.

> Fungi might contain allergens, toxins, and occasionally infectious components.
>
> Microorganisms produce large amounts of volatile microbial organic compounds (MVOCs) including alcohols, aldehydes, ketones, esters, as well as sulfur and nitric compounds. **(57)**

How does mould attach to building materials?

Given the extent of mould in the atmosphere and the serious damage it can do to the health of building occupants, it is important that more research should be done into how it attaches itself to building materials, how it varies from one material to another and what supports its growth. There has been a tendency within the construction industry to focus on damp penetration and "rising damp" over the past 50 years. The aim has been to keep water out by using materials that are water impermeable and hundreds of millions of pounds have been wasted on injecting damp-proofing chemicals into masonry and brick walls as a condition of mortgage lenders in an effort to remedy the perceived problem of rising damp. Ward describes some conventional theories of rising damp as nonsensical and points out that laboratory-based research has failed to simulate capillary action in walls in his book where he carefully investigates the movement of water through pores in building materials. He cites the work of Jeff Howell who spent much of his life investigating the apparent mythology of rising damp and the damaging effects of chemicals introduced unnecessarily into homes. **(58)**, **(59)**, **(60)**

Rising damp, if it does occur at all, is a rare phenomenon and most damp problems are usually a result of poor maintenance, broken gutters and downpipes or external earth piled up where it shouldn't be. This leads to water ingress that can provide a breeding ground for mould and dry rot but mould does not always result from penetrating damp and is more likely to be due to condensation on surfaces in buildings. Severe water penetration can lead to decay in building materials and dry rot. Increasingly modern building materials involve the use of plastics and chemicals which surprisingly can facilitate mould growth. So serious is the problem that many products like gypsum plaster, lining materials and paints are now dosed with fungicides (pesticides) in an effort to inhibit mould growth.

Some scientists argue that rising damp is a common problem in historic buildings but can be effectively treated with electromagnetic and electro-osmosis methods. A group at University of Delft investigated these claims and treatments but found they were not effective.

> Based on site field investigations and monitoring campaigns, the tests have shown that these devices did not cause any significant nor consistent decrease of the moisture content of the masonries. **(61)**

Despite clear evidence to the contrary, rising damp continues to be promoted by official bodies such as Citizens Advice as a source of damp and mould while failing to explain the significance of condensation. **(62)**

Emphasis is placed on rising damp to minimise the importance of condensation and the fact that so many modern building materials, designed to seal out moisture, provide good conditions for mould to become established. Introducing fungicides into homes seems largely unregulated despite the risks it poses to health and they do not always seem to do a very good job inhibiting mould. Fungicides are hazardous chemicals in building materials.

Literature about dealing with moisture in historic buildings tends to see the issue as one of heating and ventilation in an effort to contain relative humidity levels rather than analysing the performance of materials. **(63)**

Research on mould

While there has been a great deal of research on mycology and the effects of mould, academic research on mould and building materials has been very limited. The Centre for Moisture in Buildings at University College London (UKCMB) carries out research into this and organises a biannual conference where some useful academic papers on mould and damp can be found.

Much of the literature on mould tends to focus on temperature and general environmental conditions, with data based largely on computer simulation models, rather than field work on actual buildings. However, an important literature survey was carried out by Neil May and others calling for more research on moisture and health in buildings, and this led to the establishment of the UKCMB. This study raises the issue of "Toxic Mould," but suggests there is insufficient evidence of the effect of toxic mould on a range of serious illnesses. This study, surprisingly, has little to say about building materials, saying that more data is needed on moisture problems. One possible reason for a shyness about building materials is that the report was funded by building material manufacturer Saint Gobain. Saint Gobain is the owner of Celotex (used on Grenfell Tower) and has funded further work at the UKCMB. **(64)**

At the first international conference on Moisture in Buildings (ICMB21), UCL London 28-29 June 2021 it was announced that two research projects had begun, one of which was funded by Saint Gobain.

> We are glad to announce that two UKCMB projects have started, both of which will be presented at the upcoming conference. The first is on Health and Moisture in Buildings, funded by St Gobain Research. **(65)**
>
> The second project is on mould testing and benchmarking, sponsored by Polygon UK with the support also of Mycometer from Denmark. This project aims to set up a robust methodology for testing of mould levels in buildings and will trial this methodology on up to 100 properties to see what levels of mould are in both "dry" and "wet" buildings.

This benchmarking study was published in February 2018, funded by Polygon, a large property damage restoration company with sales of EUR 743 million in 2020. Polygon works on flood damage and asbestos detection. **(66)**

Further research by the UKCMB consisted of a survey of 33 designers and practitioners about their awareness of and experience of mould and damp (also funded by Saint Gobain). An interesting finding was that increasing energy efficiency measures have unintended consequences such as damp. There is a lot of interesting material in the appendices to the study but interestingly the word mould only appears once in the whole document.

> There appears to be widespread ignorance amongst designers, developers, clients, contractors, managers and users about how heating, ventilation and moisture work together. Single-issue treatment of carbon reduction and energy efficiency has had noticeably negative effects on moisture balance in buildings especially with respect to air-tightness. **(67), (68)**

The UKCMB has begun to draw attention to the problems of damp and mould since the survey was published, such as a study that identifies airborne "invisible" mould as well as visible mould, largely focused on developing a methodology for measuring mould as they state that mould is present in all buildings.

> The results show a clear difference between mould concentrations obtained from homes with visible mould and those without... Indoor mould problems differ from other indoor air quality issues in that mould is present in all buildings, and even non-problem properties with no building related sources of mould contain various levels of mould on the surfaces. **(69)**

Evidence of fungal and microbial effects on health in buildings

While it is obvious to most that black mould on walls, ceilings and window frames cannot be a good thing, some people are more sensitive to the resulting allergic and respiratory problems. Reducing or removing mould in homes would have enormous health benefits through reduced admissions to hospitals and an improvement in health generally. Mould can be found in schools and workplaces as well as homes but as yet there is no overall assessment of the extent of the problem in the UK. None of the projects examining indoor air in schools, funded under the UK Clean Air Programme, included mould as part of their investigations. Medical epidemiological research funded by bodies like the Medical Research council is at fault for rarely, if ever, gathering data on the conditions in homes and buildings when collecting data on health problems in children and adults.

Born in Bradford is an internationally recognised research programme which aims to find out what keeps families healthy and happy by tracking the lives of over 40,000 Bradfordians.

> We use our findings to develop new and practical ways to work with families and health professionals to improve the health and wellbeing of our communities. **(70)**

Despite the vast range of research and publications in the Born in Bradford project it was not possible to detect a single study that addresses the building and living conditions of the population taking part. Efforts to check this by arranging a meeting during a visit to Bradford were refused on the basis that leading figures in the project were too busy!

This is despite the fact that Bradford has some of the worst housing conditions in the UK with huge levels of overcrowding and poor conditions, particularly in privately rented houses and flats. **(71)**

Various projects are now underway to monitor air quality in Bradford, and in schools, though this is mostly focussed on traffic pollution. The SAMHE project, for instance, is gathering some data in schools but only CO2 and TVOCS using monitors that the school children themselves can use. Such monitors cannot identify VOC emissions from building materials or mould so cannot provide more than a general indication of problems. **(72)**, **(73)**

With a large team of 38 academics and advisors from five UK universities, Cambridge, Imperial College, York, Surrey, Leeds and the UK Health Security Agency (UKHSA), the SAMHE project is already biased towards external and emissions from cleaning products, air fresheners, rather than internal air pollution and ventilation issues. The air fresheners have been installed as part of the project.

> The monitors also detect total volatile organic compounds (TVOCs). Volatile organic compounds, or VOCs, are chemicals that evaporate into the air from liquids or solids commonly found indoors. VOCs come from a range of sources like cleaning products, air fresheners, paint or cooking. **(74)**

Reference to mould could not be found in the SAMHE project despite the serious mould problems in Bradford. **(75)**

Detailed data on VOC emissions and mould data has not been collected in Bradford, unlike in many other parts of the world where schools have to be closed for "detoxing" to remove mould or deal with other indoor air quality problems. This is all too frequent, such as ironically in *Bradford County* in Florida, USA where they have ready access to experts who can carry out proper air quality and mould testing. **(76)**

Similar work is done in various European countries with extensive publication in Finland where schools with moisture and mould problems have been compared with a control school without these problems. Multi-organ health symptoms were found in the schools with mould where Indoor air was collected with a novel condensed water sampling technique. Over 51% reported gastrointestinal problems and a wide range of other respiratory and skin problems. **(77)**

The SAMHE project is being extended from Bradford to 1000 to 2000 schools throughout the UK with additional EPSRC funding but may continue at the same superficial level rather than providing detailed data as in the Finnish studies.

Research in Denmark in the Copenhagen area has identified a wide range of bacterial species in homes though these vary with relative humidity and seasonal changes. Measurements were performed for a whole year inside five homes and once in 54 homes. The most species-rich genera were *Bacillus, Paenibacillus* and *Kocuria*. Such data is not readily available for UK homes.

Dust sampling was used rather than collection in an air pump. Some of the species detected are classified as risk class 2 pathogens, which means that they can cause human disease but are unlikely to spread to the community. They suggest that decreasing the relative humidity might be a strategy to reduce the exposure to some airborne bacterial species. Much of this work is inconclusive and requires wider study of different dwelling types and relative humidity conditions. **(78)**, **(79)**

While such data is not currently available in terms of UK homes, the team at the UKCMB say they have begun to investigate the wide range of sampling methods that are available:

> ...(at) present no testing procedure has been standardised. The vast number of different fungi species, the differences in their biological properties and their implications on the occupants' health and building fabric can make the decision-making process for an appropriate assessment protocol challenging. **(80)**

Medical research on the effects of Aspergillus and other moulds is extensive, but rarely part of a multi-disciplinary approach as medical experts rarely collaborate with building experts. Chronic pulmonary aspergillosis (CPA) and chronic pulmonary obstructive disease (COPD) appear to be linked, though COPD may have other causes than mould. Living near industrial composting sites or farms, vacuuming frequently and not asking visitors to remove shoes on home entry have been found in a study where occupants have reported greater than four antibiotics courses in a year. While these are interesting, if rather eccentric variables, there may be many more environmental factors than can be related to the effect of mould. **(81)**, **(82)**

Such medical research also discusses whether windows are open or shut, facing the sun, filtered ventilation systems, but as usual the fabric of the buildings is not considered. It is hardly surprising that academics call for more research when they fail to investigate the main cause of mould. **(83)**, **(84)**

Du et al suggest that multidisciplinary cooperation of epidemiology, building technology, as well as biological and molecular research is needed to facilitate understanding of the causal mechanism, quantitative assessment and acceptable thresholds for mould exposure and management in buildings caused by spores such as Cladosporium, Aspergillus and Penicillium. They point out a finding which is relevant in the UK and many other countries where buildings are being made more airtight.

> building energy efficient designs, such as increased insulation and airtightness improves the thermal environment but possibly accumulate indoor moisture and increase mould growth risks, due to lack of sufficient ventilation. **(85)**

Mould and building materials

It is not difficult to find bad advice and misinformation about damp and mould. For instance, an article on the Timberwise website, which claims the problem is due to the age of the British housing stock. This completely ignores the significant increase in mould in more recent buildings, whereas older breathable buildings can be less risky, until they are badly renovated and sealed up. **(86)**

Given suitable conditions the various species of mould will grow on almost any surface and building material even in new buildings. Mould seems very happy to establish on plastic and synthetic materials and it can be difficult to remove. Plasticised PVC windows for instance are commonplace and condensation on windows is one of the first signs of problems. Plasticised materials can be degraded by bacteria and fungi.

> The susceptibility of (plasticised) pPVC results from the presence of plasticizers, commonly organic acid esters such as dioctyl phthalate (DOP) and dioctyl adipate (DOA), added to modify physical or mechanical properties of the polymer. Both bacteria and fungi can degrade ester-based plasticizers. Loss of plasticizers from pPVC due to microbial degradation results in brittleness, shrinkage, and ultimately failure of the pPVC in its intended application. **(87)**

Plastic materials in buildings are often characterised as stable and safe, but in reality pvc products, which contain numerous hazardous chemicals, can slowly break down and not only provide a home for mould but also can emit a range of toxins as well as being an accelerant in fire. Dr Barbara Lane, an expert witness for the Grenfell Fire inquiry, identified badly fitted PVC windows as a contributor to the fire.

> Experts told the inquiry, the uPVC window frames appeared to have melted, exposing combustible materials plugging the cavity between the new windows and building exterior within them. **(88)**

Research into the relationship between different construction materials, surfaces and mould show that even materials referred to as natural and healthy and beneficial to the indoor environment can cause problems and be susceptible to mould. Very expensive composite cork and cork lime products have recently become very popular with ecoarchitects, but they may contain anti-fungal chemicals because of worries about mould. **(89)**

Dangers of Internal Wall Insulation (IWI) and other insulation solutions

A popular and widely used solution to damp and mould is to line internal walls with some kind of insulating board. There are a range of materials available such as calcium silicate, wood fibre and others, but the standard approach is to use plasterboard backed with a non-vapour permeable plastic foam insulation product such as PUR or PIR. Other solutions use mineral wool insulation, which is vapour permeable but may not be very resilient to moisture. Not only will the plasterboard have been treated with fungicides, but the insulation materials will be capable of off-gassing flame retardants and other toxic chemicals. As the foam degrades, the flame retardants will be released, which presents a danger when property is renovated. IWI, as it is known, is so common that it is rarely challenged, yet after a few years, mould may be found trapped behind the dry lining. In an attempt to meet higher energy standards, similar dry lining is now being used in new build construction due to the poor insulation in the main structure.

> The current UK building standards regulating mould risk, Approved Document F (ADF), do not provide clear guidance on what is to be done if a material fails its mould risk assessment criteria. In a recent study, all insulation materials used as internal wall insulation were predicted to exceed the ADF mould criteria for as high as 80% of the time. Failure within IWI can go unnoticed for long periods of time; even if recognized, it is unrealistic to replace IWI at the earliest sign of failure. **(90)**

An interesting study at Leeds Becket University has shown that IWI retrofit measures can increase the risk of mould growth in a neighbouring property where no insulation has been installed, but much of this research is based on computer modelling rather than physical observation. **(91)**

It is possible to install internal wall insulation using breathable, vapour-permeable materials.

The best solution is to use 50mm or more of hemp lime, but this is a wet material and requires time to dry out. Even though it is the best materials available, it is unpopular due to the drying time issue and prejudice about hemp. **(92)**

Some progressive retrofit bodies such as People Powered Retrofit in the North of England use dry solid wood fibre boards, which are glued or mortared onto existing walls rather than petro-chemical materials, though they are strangely shy of providing details of insulation materials on their websites. Wood fibre boards are imported from various European countries and some contain glues and additives, though are still relatively low in emissions. **(93)**

External insulation (EWI)

Problems caused by external insulation measures have already been mentioned such as in Awaab's block in Rochdale. Sealing up the outside of 19th-century brick terraced houses has also been carried out throughout the UK and continues to be promoted as a retrofit solution. This often leads to mould and damp internally, often referred to by academics as "unintended consequences." **(94)**

One of the most infamous examples is in Preston, an area called Fishwick. **(95), (96)**

This is not only a UK problem but can be found across the world. Academics in Australia have estimated that

> 1 in 3 Australian homes display excessive dampness and mould proliferation, representing a significant threat to human health. **(97)**

The US Centre for Disease control states that (American spelling of mould is mold):

> Many building materials provide suitable nutrients that encourage mold to grow. Wet cellulose materials, including paper and paper products, cardboard, ceiling tiles, wood, and wood products, are particularly conducive for the growth of some molds. Other materials such as dust, paints, wallpaper, insulation materials, drywall, carpet, fabric, and upholstery, commonly support mold growth. **(98)**

The above statement promotes an assumption that mould requires wet surfaces. It is not uncommon to find mould on dry surfaces, though there may be high levels of humidity in the room where it has appeared. However, condensation on windows leads to wet surfaces that will encourage mould. Even concrete flooring and common floor coverings such as vinyl can create mould problems. Mould attached to the floor leads to degradation and emissions of other hazards apart from mould. **(99)**

Hazardous treatment of mould

Occupants of houses suffering from mould are generally encouraged to "wash down" the mould with bleach or fungicides. These cleaners are regularly advertised on television but are also recommended by landlords private and social. It is not always made clear to occupants that concentrated chemicals such as Sodium Hypochlorite have to be diluted and also handled with protective clothing and gloves. Many concentrated bleaches and fungicides can have serious effects on health and can trigger asthmatic reactions. Furthermore, the mould can return in a week or two requiring repeated treatment. The mould spores can be embedded in the fabric and cannot always be removed with bleach. Indeed, sometimes the only way to removed embedded mould is to remove the building materials entirely, requiring significant renovation work. Sodium Hypochlorite can irritate the lungs and repeated exposure may cause bronchitis coughing, phlegm and shortness of breath. **(100)**

Materials that have been treated with "mould resistant" chemicals can include wood, insulation materials, plasterboard sometimes coated with glass fibre instead of paper, tape for gaps in plasterboard and for air tightness, caulks and sealants, spray foams, a wide range of mould resistant paints, mould resistant plasters, carpets and shower tray pans. **(101)**

Anti-mould paints are readily available from paint shops and DIY stores. Here is an example from a health and safety data sheet.

> May cause an allergic skin reaction. Harmful to aquatic life with long lasting effects. Avoid breathing vapour or spray. Wear protective gloves and eye protection: disposable vinyl gloves (EN374) and safety glasses with side-shields. Avoid release to the environment. IF ON SKIN: Wash with plenty of soap and water. If skin irritation or rash occurs: Get medical attention.

The paint is made of zinc oxide, 4,5-dichloro-2-octyl-2Hisothiazol-3-one and bronopol (INN). Isothiazolone which has high acute inhalation toxicity and may produce an allergic reaction. **(102)**

As mould has become so common in houses, the sales of hazardous chemical treatments have increased, but this is only introducing more volatile chemicals into the home and work environment. Various less toxic materials are available for washing or brushing off mould such as a solution of baking soda or vinegar. Often referred to as old wives' tales, these materials can be effective, though possibly less so once mould is very well established.

The UK Health and Safety Executive publish a list of approved biocidal products which includes details of the chemical constituents and whether it can be used by amateurs or professionals or both. It includes amateur risk mitigation measures, but there is no indication of whether they take any view on its safety. The list of thousands of products includes dog flea treatment, boat anti-fouling as well as mould treatment! **(103)**

Electrical appliances such as fridges and freezers are now coated with biocides, and come with leaflets saying this is not hazardous to health, but it could contaminate food stored.

Keeping the rain out

It has been common practice for many years to assume that condensation and mould are a result of water penetration. The construction industry has been obsessed with sealing up buildings to keep water out. Older properties have been rendered with cement-based materials such as pebble dash and a wide range of waterproofing materials. As the retrofit movement has developed, external insulation materials have been used, usually covered with a polymer cement, acrylic and silicone render, to keep the rain out. The range of products range from thin to thick coats, plastered or sprayed and are often referred to in the promotional literature as *organic*, in an effort to make them sound natural.

Health and safety data sheets for a wide range of render systems will show that they contain substances such as "Distillates (petroleum), hydrotreated heavy naphthenic; 2-methylisothiazol-3(2H)-one, Titanium dioxide, Zinc Pyrithione" (source not disclosed for legal reasons). It is clearly stated that these chemicals can cause allergic skin conditions and cancer.

Acrylic and other similar renders systems are not vapour permeable, though proprietary systems are often advertised as breathable and so hundreds of thousands of houses have had their walls sealed up by these materials. Initially the houses look smarter and better, but there are questions as to how long these renders will last. The failure of some systems can be detected through the disappearance of well-known brand names over recent years and failures are settled out of court by insurers, making it hard to understand what has occurred. Polymer render systems have been introduced in the belief that they are less likely to fail than traditional cement and sand. Various layers may be included such as a fibreglass mesh and there is a range of so-called "breathable" renders available. One external wall insulation render system is made from expanded polystyrene (EPS) and claims to be breathable. Needless to say, EPS is not vapour permeable (breathable) so one company drills hundreds of little holes 2–3 mm wide into the polystyrene! The final render coat also is treated with chemicals against mould, and adhesives are also part of the system

Warranty companies will admit that render failures are a major source of claims. **(104)**, **(105)** Waterproofing "creams" are widely used to keep water out of brick and masonry buildings.

One well-known brand, often marketed as "magic," contains the following chemicals: Phenol, methylstyrenated, fatty acids, c18-unsatd., dimers, polymers with tall-oil fatty acids and triethylenetetramine benzyl alcohol, formaldehyde, polymer with benzenamine, hydrogenated xylene, 2,4,6-tris dimethylaminomethyl, phenol ethylbenzene and salicylic acid. Fortunately, this chemical cocktail is on the outside of buildings, but this does not mean that emissions cannot migrate to the inside. A less toxic silane alternative includes Triethoxy (2,4,4-trimethylpentyl), silane, Triethoxyoctylsilane, alkanes and C11 – C15 (iso) Octyl silane hydrolysate.

Such treatment is recommended in an academic research paper as a solution to avoid long-term moisture build up. The authors claim that using a waterproofing cream externally allows a "damp wall" to dry without significant loss of vapour permeability. This was used in a project that had been retrofitted with internal wall insulation, using a polyurethane spray foam, with the implication that this maintained vapour permeability (as some spray foam products claim their insulation is "vapour open"). This paper contains a great deal of technical detail, particularly as to how the moisture levels were monitored, and while a computer modelling system was used, data loggers to obtain on-site information were also part of the study. The authors do not refer to the chemical constituents of the materials used, and their possible effect on human health or the environment, and they also declare that there were no conflicts of interest even though two of the authors worked for Safeguard, the company supplying the waterproofing cream! **(106)**

The great retrofit disaster

There has been a great deal of talk about the importance of retrofitting houses and buildings. This is presented in various ways as "decarbonising" and part of Government net zero carbon strategies. Various grant schemes in England, Wales and Scotland such as "ECO" (the latest is ECO4) have been introduced by the Government but uptake has been very poor among people suffering from fuel poverty whereas middle class people are more likely to benefit if they can navigate the grant conditions. **(107)**

The retrofit measures are poorly thought out with the most recent ECO 4 delivery guidance failing to discuss insulation materials, while suggesting cavity, external and internal insulation. The grant-funded work (normally worth only £7,000) is largely insufficient to put houses in a good state of repair or even improve their energy efficiency, but people on low income cannot afford the additional work to resolve all the problems found in many older dwellings. And so the uptake of grants is very poor.

Grants for external wall insulation (EWI) can be worth up to £14,000. **(108)**

Some retrofit measures can lead to poorer indoor air quality and often to mould and damp. Many of retrofit "experts" and agencies responsible for grant-funded and retrofit schemes seem unaware of these indoor air quality problems. Indeed, the materials and methods which they are encouraged to use are often unsuitable and inappropriate. This has been demonstrated in numerous reports and has been referred to over the past decade by journalists such as De Selincourt as the "retrofit disaster" problem. **(109)**

Training of approved retrofit installers, retrofit co-ordinators and the introduction of British Standards (PAS2035) for retrofit have barely touched on the retrofit disaster problem and so mistakes continue to be made. When a householder is considered eligible for a grant a surveyor will visit the property. The grant funding comes from electricity supply companies and many of the websites advertising grants are focussed on installing electric heat pumps rather than adopting a holistic approach to necessary renovation. **(110)**

Essentially there are four main problems:

1. Insulation measures usually involve non-vapour permeable plastic insulation materials. Polystyrene is the most common as it is the cheapest but PIR and PUR, isocyanate-based materials are often used
2. Insulation on the inside or the outside of buildings is then further sealed up with polymer and cement-based render systems which reduce the vapour permeability even further. Even where permeable insulations are used such as mineral wool, the walls are still sealed up with cement renders and gypsum plasters.
3. Insufficient ventilation is provided once buildings have been sealed up. They might have been draughty and leaky but regulations for adequate ventilation are insufficient.
4. A wide range of chemical emitting materials are used in retrofit projects with no attention paid to assessing indoor air quality.

External wall insulation (EWI)

Houses with solid walls (mainly brick) do not have a cavity that can be filled with insulation (even though this is not a good idea). Instead, it is assumed that the best approach is to over-clad the walls with insulation render systems. This approach can lead to serious problems in which houses become full of damp and mould and many EWI measures have been removed on high-rise and even medium-rise buildings since Grenfell, as the use of external polystyrene and other plastic foam materials have been seen as a fire hazard.

Fishwick Preston

As there are literally thousands of EWI retrofit disasters around the UK, this problem is best illustrated by the best-known example in Fishwick, Preston Lancashire. Fishwick was a relatively attractive area of two-storey red brick terraced houses in the east of Preston. The streets ran off a busy main road with plenty of shops with a strong local community. Fishwick was chosen as a demonstration of a government initiative called Community Energy Savings Programme (CESP). Beginning in 2009, renovation work was carried out to approximately 400 houses in the area with a spending of public money which ran into millions. As the houses were solid brick construction, 387 houses received external wall insulation (EWI) using a method involving plastic non-breathable insulation and cement render coating. The funding came from an electricity company (Intergen), managed by ANESCO with an EWI installer, Ecogen, using an approved system made by Wetherby. The EWI work was carried out in 2013 using polystyrene boards covered with a cement pebble dash, PVC nosings and corner beads and roof and eaves capping. **(111), (112)**

The problems caused by sealing up the buildings became apparent very quickly and mould and extensive damp began to appear in houses that had previously had no such problems. This was well documented by journalist Andrea de Selincourt. **(113)**

The EWI contractor Ecogen went into liquidation almost immediately in October 2013. **(114)**

It's not clear what happened to Anesco, but its director is listed as having resigned from directorships of 28 other companies. **(115)**

Dozens of similar companies around the UK cashed in on Government retrofit schemes and then disappeared almost as soon as they had collected the funding, despite offering generous guarantees. The Government energy regulator, OFGEM, began an investigation in March 2014 and served an enforcement order for 62 houses to be remedied. The electricity company, Intergen, carried out what they claimed to be independent surveys in 2015, and the EWI system supplier, Wetherby, claimed there were only a few cases of water ingress. (This is not surprising as the problem was largely one of condensation and trapped moisture.) Andrea Howe the council manager listed far more damp and mould problems and even collapsed ceilings. Residents began to develop serious health problems as a result.(Wetherby continue to supply EWI in sulation and render systems). **(116)**

The massive failure was attributed to incorrect specification, a lack of skilled workers, managing agents and contractors with lack of knowledge, profit maximisation and lack of on-site supervision, while non one thought to blame the material technology itself. Figures had been published claiming massive carbon savings from the scheme which OFGEM had to nullify.

Having wasted millions of pounds of Government grant money, a further £1.5 million was allocated to carry out works to the 62 properties subject to an enforcement order, but Preston council claimed that 390 houses needed to be fixed. Householders (tenants and owner occupiers tried to claim on their house insurance but the insurers refused to pay out unless all the EWI was removed. Years passed by with occupants left in appalling conditions while the bureaucrats argued about what to do and how to fund the renovations. The fuel poverty charity National Energy Action stepped in and agreed to renovate a small number of houses of those people "in greatest need" and began to talk to the local community in 2020. The project was slowed down due to COVID. **(117)**

Published information on the NEA project provides little detail and the staff involved seem to come and go very rapidly but some information was obtained from NEA staff by email. NEA said they would remove the existing EWI and allow the property to dry out, though it wasn't clear whether the occupants would remain in residence, and then repair the damage that had been caused. NEA stated that the earlier defects were due to poor and rushed workmanship and poor detailing and that it had been installed during a very wet period. Weather archives for the construction period do not show out of the ordinary wet weather at this time, but it was apparent that NEA did not want to blame the EWI system itself. Despite the extensive review of the problems in Fishwick (including a visit by this author) the lessons were not learned that installing non-breathable insulation systems had led to the problems.

What is even more remarkable is that NEA, according to Lorraine Donaldson, Head of Major Projects at NEA, decided to install another expanded polystyrene EWI system with a silicone render finish, this time supplied by K Systems. **(118)**

The trade literature states that the silicone finish allows water vapour to pass through and that the material contains a biocide to prevent mould infection. Some polystyrene manufacturers claim that EPS insulation containing graphite is breathable. **(119)**

Here was an opportunity to use healthy breathable materials but NEA decided to seal up the buildings with more plastic, as was done in the initial disaster. Time will tell whether this leads to yet more problems.

Cavity wall insulation (CWI)

When retrofitting housing began, cavity wall insulation was seen as the best way to make older houses more energy efficient. Cavity walls were introduced in the early part of the 20th century to provide an air gap (air is a good insulator) and the theory was that any water ingress through the outer leaf (usually brick) would be allowed to run down the cavity and escape through weep holes at the bottom of the wall. Filling the cavity with glass fibre or polystyrene beads and even urea formaldehyde could easily block the cavity, allow moisture to build up and once wet, provide little insulation value. Hundreds of thousands of houses became contaminated with mould and damp as a result of CWI. Campaign groups like CIVALLI were set up and have documented hundreds of cases of people whose houses and lives were ruined by filling up cavities. **(120)**

Despite a CWI guarantee scheme, the warranty providers often refused to pay out and people found their houses could not be sold or remortgaged. It can be argued that CWI has played the biggest role in creating an epidemic of mould and poor indoor air quality. **(121)**

Large numbers of CWI installers went bankrupt, whereas others shut down and re-emerged as cavity wall insulation extractors to remove the material they had previously installed. Some local authorities are extracting CWI from hundreds of houses. A study by the Northern Ireland Housing Executive of a thousand houses that had received CWI installation found that 63% of the houses were defective. **(122)**

Despite the extensive evidence of the failure of CWI, it is still promoted in 2023 Government house improvement grant schemes like ECO4.

Internal wall insulation (IWI)

Some commonly used internal wall insulation materials and systems present one of the biggest threats to indoor air quality and the health of occupants.

> Internal wall insulation can cause condensation if not installed properly. Condensation occurs when warm, moist air comes into contact with a cold surface and condenses into water droplets. If the insulation is not installed with proper ventilation and airtight measures, it can trap moisture within the cavity wall insulation, leading to condensation and mould growth. This can cause structural damage to the building and lead to health problems for the occupants. **(123)**

Many social housing landlords are now installing IWI in damp and mouldy houses. As suggested in the above quotation from a commercial installer, making houses airtight will only make matters worse. Conventional IWI usually involves dry lining walls with sheets of gypsum board bonded to a synthetic foam insulation batt. Such a system is not normally vapour permeable and so interstitial condensation can occur behind the dry lining and after a few years the hidden mould will build up. Installing materials dosed with fungicides, flame retardants and other chemicals on the inside of houses is one of the most irresponsible thing being encouraged by the Government, landlords and professional retrofit "experts." Not only is dry lining of this type being used to retrofit houses, but it is used in modern construction to boost insulation levels as there is usually insufficient insulation in the main walls. The majority of modern houses currently being constructed use this form of insulation. Internal finishing with plasterboard is popular with houses builders as it reduces normal plastering, with a jointing compound and tape, which is simply painted over, usually with non-breathable plastic paints.

Many architects, surveyors and builders have little expertise in renovating buildings. They generally have more experience with new builds and so tend to use the same materials and techniques in renovation. This is a big mistake and so many renovated buildings can end up worse than they were before any work is done. Sealing up buildings with synthetic insulations and plastic membranes can lead to a wide range of serious health problems and there is plenty of evidence of this. Botched renovation is not just a question of bad builders and poor training (though these are major problems), but the materials used are frequently not suitable. There can be indoor air problems in existing buildings, but these can be considerably worse after renovation.

This may be the biggest cause of indoor air health problems as most people live in existing older buildings, so while it is possible to construct healthy new homes (though this is rarely done) much of the current housing stock will be with us for many years.

Insulating buildings so they are warmer and use less fuel seems an obvious thing to do particularly as heating and electricity bills are going up in price at an astronomical rate. Fuel poverty is becoming an even greater problem and many people cannot afford to keep warm in the winter. There is a large academic community of social scientists working on fuel poverty with many thousands of research papers and books, but this has negligible impact on the daily lives of people in cold damp houses. There are many claims that insulating homes leads to houses being warmer and this ensures that people are also healthier. These claims are made frequently to justify more research funding, but in fact the evidence that people are healthier in better insulated homes is rather flimsy. It would be great to believe that there is a simple correlation between better insulation and health but the evidence simply isn't there. **(124)**

Hygroscopic materials

Conventional retrofit measures tend to use plastic non-breathable materials and seal in the damp. It is possible to use vapour-permeable materials that reduce the mould and damp risks, but this is not promoted by energy agencies, government and energy saving trust advice. Some vapour-permeable materials such as wood fibre, cork, clay, hemp and hemp lime are not only breathable but they are hygroscopic. This means that they can regulate or "buffer" humidity inside buildings, maintaining a suitable relative humidity of around 55% even when the ambient air internally or externally has high humidity. **(125)**

The evidence of the different performance of hygroscopic materials has been investigated by Latif and others using an environmental chamber. The performance of mineral wool (where a vapour barrier had to be used) showed that the higher fluctuation of vapour pressure was the lowest "due to the low thermal and negligible inertia of mineral wool." Latif has also demonstrated how moisture buffering takes place and was better when clay-based paint was used to finish the wall buildups. **(126), (127)**

Walker has compared the ability of different insulation materials to deal with moisture in a historic building in Dublin. Different layers included Lime Plaster, Aerogel, Cork Lime, Hemp Lime, Calcium Silicate Board and Thin PIR (polyisocyanurate) board. After one year, the least vapour-permeable and capillary active insulation, PIR and aerogel had higher moisture contents than the lime-based insulation (cork and hemp). There is a great deal of useful information in this study about thermal as well as moisture performance. **(128)**

An understanding of building physics in terms of moisture in buildings is still at a low level in the UK. De Selincourt, normally a very good journalist, promotes the myth that open fireplaces were good to keep houses healthy. In reality open fireplaces drew in cold air and may well only removed damp air when a blazing fire was lit. However, it is likely that sealing up chimneys has contributed to mould and damp problems.

> In the past, if traditional buildings achieved a 'healthy' moisture and ventilation performance, open fireplaces are likely to have played a big role. A constant, vigorous draught (all year round, but especially when the fire was lit), will have removed far more moisture than would ever have passed from inside to outside though a 'breathing' fabric. Now that the open fireplace is generally a thing of the past (decades past, in most dwellings), alternative moisture removal strategies are essential. **(129)**

De Selincourt appears to promote the benefits of hygroscopic materials but gives a misleading impression that vapour open materials are letting moisture in. It's a common error to confuse vapour permeability with water ingress.

> The value of hygroscopic surfaces in buffering hour-to-hour fluctuations in indoor moisture is a different issue – but hygroscopic surfaces can't remove moisture either. Hygroscopicity can also have a role in dealing with dampness that penetrates into the fabric. However, it is often hygroscopic and/or vapour open materials that are letting the moisture in, in the first place. Traditional building materials don't automatically solve all damp problems!

The petrochemical insulation companies like to play down the value of even existence of vapour permeability. As study by Cambridge Architecture Research (CAR) commissioned by Kingspan Insulation claimed that,

> ...vapour diffusion (Breathability) does not make a significant contribution to the rate of vapour transfer for a house. ... .the breathability of insulation products are therefore at best a side show in reality they are a complete red herring in the avoidance of surface condensation, mould growth and exacerbated dust mite populations. **(130)**

The claims of Kingspan, in what became known as the Red Herring report, were roundly attacked in a series of critiques and even the main CAR report was ambiguous in its evidence. However, it is clear that the petrochemical insulation industry felt threatened by the growth of natural, vapour open and hygroscopic materials as their products could not provide such performance and confusion about building physics and moisture still remains among many building and retrofit practitioners.

Retrofit insulation

It's possible to insulate homes when retrofitting, but it needs to be done with the correct material that does not make the indoor air quality worse. Using vapour-permeable and hygroscopic materials can significantly reduce the risk of damp and mould. Despite this, it is clear that the vast majority of academics and experts who talk about retrofit have very little interest in the details of how houses are insulated or retrofitted. The word insulation is intoned as a kind of mantra without any attempt to distinguish between different kinds of insulation and how they are installed. It is not good enough simply to state that houses should be retrofitted without explaining what kind of insulation measures should be taken.

Government advice on insulation by Sarah Price of Enhabit includes some questionable building physics that gives designers leeway to use vapour-closed solutions.

> The retrofit assessment should be sufficient for the designer to conclude what the existing moisture strategy is. Not all solid walls will be moisture open and indeed many have already been altered since they were built and may now include vapour impermeable layers and could now be moisture closed. **(131)**

It is common that over the years, the permeability of many old walls has been damaged by the application of cement and pebble dash renders, but this is not an excuse to seal up the walls even more with non-breathable insulation materials. Price et al go on to suggest that gaps behind insulation should be minimised by:

> Using a liberal amount of adhesive to secure the insulation to the wall. … .and the new guidance clarifies that there may be other situations where moisture-closed insulation may be suitable, provided the Retrofit Designer can demonstrate that the risk is acceptable by means of a moisture risk assessment or reference to BS5250:2021.

The Retrofit academy refers to "Trust Mark" guidance based on the Sarah Price work and continues to advocate "vapour open" insulation but without ever explaining what materials might meet these criteria. Advice on achieving good indoor air quality may be included in the Retrofit Academy guidance and training, but could not be found. The British Standards Institute PAS 2035 and related standards, currently being updated, incorporate the Trust Mark concept. While this refers to indoor air quality Trust Mark does not provide any information as to what this means. **(132), (133), (134)**

Enhabit (Sarah Price) who wrote this flawed Government guidance on retrofit practice is now part of a business called Efficient Building Solutions, which seems closely linked to commercial building materials material suppler Green Building Store. Enhabit may no longer exist and Sarah Price is now at an organisation called QUODA. Both organisations are closely linked to the passivhaus movement and so cannot be seen as independent of commercially vested interests. **(135), (136), (137), (138)**

A Scottish study of indoor air quality resulting from modern airtight construction advocates the use of materials to regulate humidity:

> It is possible to use a technique known as 'passive humidity buffering' to reduce fluctuations in humidity. This technique relies on hygroscopic materials – which readily absorb and desorb (release) moisture from the air – as surface materials thereby acting as a buffer. Clay plasters are among the best products for this but other, largely natural materials will also work.
>
> However, the most common way to manage humidity is to ventilate, either mechanically or naturally. Mechanical systems can be controlled using humidistats which automatically boost the system when excess humidity levels are sensed. Since incoming external air is almost always drier and cooler than internal air, then ventilating a property will lower the aggregate temperature and humidity of the air in the building as long as it is adequately mixed. **(139)**

Given the poor understanding of building materials and their role in indoor air quality and the development of mould, we are likely to have to live with increasing levels of mould and the resulting health problems for some years to come. Even young radical green campaigners have been seduced by passivhaus and the use of petrochemical retrofit solutions, though some are embracing the concept of vapour-permeable solutions.

Notes

1 From the content of CIH training:
What will I learn? Ethical practice in housing ‡ Understand ethical practice ‡ Understand how ethical practices apply to housing organisations ‡ Understand how ethical practices apply to a housing professional. Professional practice skills for housing ‡ Understand the concepts of being a member of a profession and acting professionally ‡ Understand the skills required to be a housing professional ‡ Be able to assess own professional performance ‡ Be able to manage own professional development Strategic and business planning for housing ‡ Understand the role, purpose and complexity of a housing organisation ‡ Understand theories to enable strategy development in a housing organisation ‡ Be able to apply strategic planning techniques to develop a business plan for a housing organisation ‡ Understand how to implement and monitor a business plan Leadership and management in housing ‡ Understand the relationship between management and leadership ‡ Understand how leadership styles impact on the achievement of organisational objectives ‡ Understand the skills and attributes required to be an effective manager Housing in context ‡ Be able to undertake a comparative study of one aspect of housing policy ‡ Understand the social, cultural and historical development of one aspect of housing policy ‡ Understand the political and economic context of housing with particular regard to one aspect of housing policy ‡ Understand the law in relation to one aspect of housing policy ‡ Be able to make a case for future trends in one aspect of housing policy Managing relationships in housing ‡ Understand how the changing profile of housing customers impacts on the delivery of housing services ‡ Understand how to communicate with housing customers ‡ Understand the variety of partners involved in the delivery of housing services ‡ Understand the principles of contract management in the delivery of a housing service

2 The Expert panel that drew up the Mould and Damp Guidance published on the website.

"The guidance was developed with an expert advisory group, with individuals from the following organisations:" op cit https://www.gov.uk/government/publications/damp-and-mould-understanding-and-addressing-the-health-risks-for-rented-housing-providers/understanding-and-addressing-the-health-risks-of-damp-and-mould-in-the-home--2#:~:text=As%20this%20guidance%20also%20makes,structural%20issues%20or%20inadequate%20ventilation.
• Dr Marcella Ucci, University College London (UCL)
• Dr Hector Altamirano-Medina, UCL Institute for Environmental Design and Engineering
• Dr Hywel Davies, Chartered Institution of Building Services Engineers
• Professor Tim Sharpe, University of Strathclyde
• Dr Jennifer Townshend, Newcastle Upon Tyne Hospitals NHS Foundation Trust
• Ian Sanders, Hull City Council and Chartered Institute of Environmental Health Board
• Annie Owens, National Housing Federation

• Sarah Davis, Chartered Institute of Housing
• Laura Wharton, Wigan Council
• Gavin Dick, National Residential Landlords Association
• Chloe Fletcher, National Federation of Arm's Length Management Organisations
• Paul Price, Association of Retained Council Housing
• Caritas Charles, Tenant Participation Advisory Service
• Dr Henry Dawson, Cardiff Metropolitan University
• Deborah Garvie, Shelter
• Sally Morshead, Shelter
• Dr Jill Stewart, University of Greenwich

References

(1) Crummy, H. (1992). Let The People Sing. Available at: www.andrewcrummy.com/let-the-people-sing [Accessed 24 Sep. 2023].

(2) Gourlay, K. (2022). 'Latest phase of Edinburgh housing 'masterplan' would deliver 150 new homes.' *Edinburgh Live*. Available at: www.edinburghlive.co.uk/news/edinburgh-news/latest-phase-edinburgh-housing-masterplan-25296922 [Accessed 24 Sep. 2023].

(3) Farr, J. and Wood, H. M. (2023). 'Scots mum dies just weeks after daughter complained about horrific mould in flat.' *Daily Record*. Available at: www.dailyrecord.co.uk/news/scottish-news/scots-mum-dies-kidney-failure-29834810 [Accessed 24 Sep. 2023].

(4) Wates, N. and Knevitt, C. (1987). *Community Architecture: How People Are Creating Their Own Environment*. Abingdon, Oxfordshire: Routledge.

(5) Spatial Agency. (1978-85). Community Technical Aid Centres. Available at: www.spatialagency.net/database/community.technical.aid.centres [Accessed 24 Sep. 2023].

(6) Palli, T. L. (2011). Housing rents and the tenants struggle in Scotland-Solidarity. Available at: https://libcom.org/library/housing-rents-tenants-struggle-scotland [Accessed 24 Sep. 2023].

(7) Bryant, R. (1979). *The Dampness Monster: A Report of the Gorbals Anti-Dampness Campaign*. Scottish Council of Social Service.

(8) Wright, V. (2021). 'housing problems … are political dynamite': Housing disputes in glasgow c. 1971 to the present Day. *Sociological Research Online*, 26(4), 976–988. Available at: https://doi.org/10.1177/1360780418780038 [Accessed 24 Sep. 2023].

(9) Bishopsgate Institute. (n.d). *Tower Blocks UK Archive*. Available at: www.bishopsgate.org.uk/collections/tower-blocks-uk-archive [Accessed 24 Sep. 2023].

(10) Glendinning, M. and Muthesius, S. (1994). *Tower block: modern public housing in England, Scotland, Wales, and Northern Ireland*. New Haven: Published for The Paul Mellon Centre For Studies In British Art By Yale University Press.

(11) Markus, T.A. and Nelson, I. (1985). An Investigation of Condensation Dampness at Darnley January 1985, Glasgow District Council (not available on-lime).

(12) Clements, C. (2023). 'Government insulation scheme ruined my home.' *BBC News*. Available at: www.bbc.co.uk/news/uk-scotland-glasgow-west-65543897 [Accessed 24 Sep. 2023].

(13) Apps, P. (2022). *Show Me the Bodies*. London: Oneworld Publications.

(14) Berrill, L. (2023). 'Clarion Housing under fire over Basildon mouldy flat.' *Echo*. Available at: www.echo-news.co.uk/news/23266070.clarion-housing-fire-basildon-mouldy-flat/ [Accessed 24 Sep. 2023].

(15) Anderson, N. (2022). 'Britain's 'mould epidemic' made worse by cost-of-living crisis: How 4.7million private renters have battled fungus in their homes over the past year as hard-pressed tenants cut back on heating.' *Daily Mail.* Available at: www.dailymail.co.uk/news/article-11456351/Britains-mould-epidemic-worse-cost-living-crisis [Accessed 24 Sep. 2023].

(16) Imperial Cleaning. (n.d). Available at: https://imperialcleaning.co.za/mold-removal/#:~:text=Mold%20Removal%20Services&text=We%20are%20the%20experts%20at,3432%20for%20a%20free%20Quote [Accessed 24 Sep. 2023].

(17) Booth R. Awaab Ishak Death (2022). Available at: www.theguardian.com/society/2022/dec/15/awaab-ishak-death-rochdale-housing-chair-to-quit-after-damning-report

(18) Pierce, G. (2015) 'Hyssop, Cedar, and Scarlet' *blogas.* Available at: www.blogos.org/exploringtheword/hyssop-cedar-scarlet.php [Accessed 24 Sep. 2023].

(19) Courts And Tribunals Judiciary. (2022) Awaab Ishak: Prevention of future deaths report. Available at: www.judiciary.uk/prevention-of-future-death-reports/awaab-ishak-prevention-of-future-deaths-report/ [Accessed 24 Sep. 2023].

(20) Annex A: Regulation 28: Report to Prevent Future Deaths (1). [pdf]. Available at: www.judiciary.uk/wp-content/uploads/2021/04/Ella-Kissi-Debrah-2021-0113-1.pdf [Accessed 24 Sep. 2023].

(21) Booth, R. (2022). 'Grenfell fire: focus shifts to possible criminal convictions as inquiry ends.' *The Guardian.* Available at: www.theguardian.com/uk-news/2022/nov/10/grenfell-fire-focus-shifts-to-possible-criminal-convictions-as-inquiry-ends [Accessed 24 Sep. 2023].

(22) BBC NEWS (2023) 'Luke Brooks: State of mouldy home did not cause death – inquest.' Available at: www.bbc.co.uk/news/uk-england-manchester-66442110 [Accessed 24 Sep. 2023].

(23) Holmes, W. (2023) 'Mum died after moving to mouldy house that caused black dirt to come out of her nose.' *Echo.* Available at: www.liverpoolecho.co.uk/news/liverpool-news/mum-died-after-moving-mouldy-27155135 [Accessed 24 Sep. 2023].

(24) Booth, R. (2023). 'Social landlord in England said mould was "acceptable" in refugees' homes' *The Guardian.* Available at: www.theguardian.com/world/2023/mar/28/social-landlord-england-said-mould-acceptable-refugees-homes [Accessed 24 Sep. 2023].

(25) Hewitt, D. (2021). 'ITV News investigation exposes racism in social housing sector.' *ITVX.* Available at: www.itv.com/news/2021-11-09/itv-news-investigation-exposes-racism-in-social-housing-sector [Accessed 24 Sep. 2023].

(26) Kearns, A., Whitley, E. and Curl, A. (2019). Occupant behaviour as a fourth driver of fuel poverty (aka warmth & energy deprivation). *Energy Policy*, [online] 129, 1143–1155. Available at: https://doi.org/10.1016/j.enpol.2019.03.023 [Accessed 24 Sep. 2023].

(27) Cayla, J.-M., Maizi, N. and Marchand, C. (2011). The role of income in energy consumption behaviour: Evidence from French households data. *Energy Policy*, 39(12), 7874–7883. Available at: ttps://doi.org/10.1016/j.enpol.2011.09.036 [Accessed 24 Sep. 2023].

(28) Blight, T.S. and Coley, D.A. (2013). Sensitivity analysis of the effect of occupant behaviour on the energy consumption of passive house dwellings. *Energy and Buildings*, 66, 183–192. Available at: https://doi.org/10.1016/j.enbuild.2013.06.030 [Accessed 24 Sep. 2023].

(29) Monahan, S., Gemmell, A. (2012). The Impact of Occupant Behaviour and Use of Controls on Domestic Energy Use, *NHBC Foundation NF38.* Available at: www.nhbcfoundation.org/Researchpublications/tabid/339/Default.aspx [Accessed 26 Sep. 2023].

(30) Yohannis, Y.G. (2012). 'Domestic energy use and householders' energy behaviour', *Energy Policy*, 41, 654–665. Available at: http://linkinghub.elsevier.com/retrieve/pii/S0301421511009050 [Accessed 22 Oct. 2013]. [Accessed 26 Sep. 2023].

(31) Palmer, J. (2018) 'What are the Barriers to Retrofit in Social Housing? Report for the Department for Business, Energy and Industrial Strategy.' [pdf] *Cambridge Architecture Research.* Available at: https://assets.publishing.service.gov.uk/government/uploads/system/uploads/attachment_data/file/787361/Barrier__to_Retrofit_in_Social_Housing.pdf [Accessed 26 Sep. 2023].

(32) Baker, K.J., and Mould, R. (2020). 'Evidence Summary on Energy Performance Certificates and the delivery of energy efficiency and fuel poverty policies in Scotland' (Redacted version published 04/

10/21). Available at: https://drive.google.com/file/d/1fLQK86epIiQuDDzN9ObyZbbDMzZP5bHt/view?usp=sharing [Accessed 26 Sep. 2023].
(33) Mould, R., and Baker, K.J. (2017). 'Uncovering hidden geographies and socio-economic influences on fuel poverty using household fuel spend data: A meso-scale study in Scotland', *Indoor and Built Environment*, 26(7), 914–936. Available at: https://doi.org/10.1177/1420326X17707326 [Accessed 26 Sep. 2023].
(34) Mould, R., and Baker, K.J. (2017). 'Documenting fuel poverty from the householders' perspective', *Energy Research and Social Science,* 31, pp. 21–31. Available at: https://doi.org/10.1016/j.erss.2017.06.004 [Accessed 26 Sep. 2023].
(35) Courts and Tribunals (2022). Op cit.
(36) Richardson, M. and Rautemaa-Richardson, R., (2021) 'Aspergillus in indoor environments'. In *Encyclopaedia of Fungi.*
(37) Courts and Tribunals (2022). Op cit.
(38) FOI response from Housing Ombudsman to author (202226650) (24.2.23).
(39) Housing Ombudsman Service. (2021). Spotlight on damp and mould. [pdf] Available at: www.housing-ombudsman.org.uk/wp-content/uploads/2021/10/Spotlight-report-Damp-and-mould-final.pdf [Accessed 26 Sep. 2023].
(40) The Ombudsman also provided information on recent damp and mould data revealing the massive increase in a recognition of the problem.
(41) National Housing Federation. (2023). Regulator of Social Housing report published on damp and mould. Available at: www.housing.org.uk/news-and-blogs/news/regulator-of-social-housing-damp-and-mould-report/#:~:text=The%20Regulator%20of%20Social%20Housing,from%20others%20in%20the%20sector [Accessed 26 Sep. 2023].
(42) Apps P. (2022). Op. cit.
(43) Topping, S., and Sommerlad, N. (2022) 'Everyone has had enough on the estate where it took a child's death for the country to listen.' *Manchester Evening News.* Available at: www.manchestereveningnews.co.uk/news/greater-manchester-news/everyone-enough-estate-took-childs-25533637 [Accessed 25 Feb. 2023].
(44) Dunkerley, S. (2023). Email to author, 23 February.
(45) McCarthy, D. (2022). 'Responding to the Regulator in respect of damp and mould – Five key actions for legal compliance.' Available at: www.cih.org/blogs-and-articles/responding-to-the-regulator-in-respect-of-damp-and-mould-five-key-actions-for-legal-compliance. [Accessed 26 Sep. 2023].
(46) CIH. (2023). Available at: www.cih.org [Accessed 26 Sep. 2023].
(47) Gov. UK (2023). Social housing managers must be qualified under new laws to protect residents. Available at: www.gov.uk/government/news/social-housing-managers-must-be-qualified-under-new-laws-to-protect-residents [Accessed 26 Sep. 2023].
(48) Johansson, S., Wadsö, L., and Sandin, K. (2010). Estimation of mould growth levels on rendered façades based on surface relative humidity and surface temperature measurements. *Building and Environment*, 45(5), 1153–1160. Available at: https://doi.org/10.1016/j.buildenv.2009.10.022 [Accessed 26 Sep. 2023].
(49) AQEG. (2022). Report: Indoor Air Quality. Available at: https://uk-air.defra.gov.uk/library/reports.php?report_id=1101 [Accessed 26 Sep. 2023].
(50) Heritage House. (n.d). The Fraud of Rising Damp. Available at: www.heritage-house.org/damp-and-condensation/the-fraud-of-rising-damp.html [Accessed 26 Sep. 2023].
(51) Gov. UK (2023). Understanding and addressing the health risks of damp and mould in the home. Available at: www.gov.uk/government/publications/damp-and-mould-understanding-and-addressing-the-health-risks-for-rented-housing-providers/understanding-and-addressing-the-health-risks-of-damp-and-mould-in-the-home--2#:~:text=As%20this%20guidance%20also%20makes,structural%20issues%20or%20inadequate%20ventilation [Accessed 26 Sep. 2023].
(52) Richardson (2021). Op cit.

(53) CDC. (2022). Basic Facts about Mold and Dampness. Available at: www.cdc.gov/mold/faqs.htm#:~:text=Many%20building%20materials%20provide%20suitable,the%20growth%20of%20some%20molds [Accessed 26 Sep. 2023].
(54) Hyvönen, S. (2020). 'Association of toxic indoor air with multi-organ symptoms in pupils attending a moisture-damaged school in Finland.' *Am J Clin Exp Immunol* 9(5), pp.101–113
(55) Norback, D. et al. (2011). 'Lung function decline in relation to mould and dampness in the home: the longitudinal European Community Respiratory Health Survey ECRHS II. *Thorax*, 66(5), pp.396–401. Available at: https://doi.org/10.1136/thx.2010.146613 [Accessed 26 Sep. 2023].
(56) Wieslander, G. (2010). 'Dampness and 2-Ethyl-1-hexanol in Floor Construction of Rehabilitation Center: Health Effects in Staff.' *Archives of Environmental & Occupational Health*, 65(1), pp.3–11. Available at: https://doi.org/10.1080/19338240903390248 [Accessed 26 Sep. 2023].
(57) Mousavi, B. (2016). '*Aspergillus* species in indoor environments and their possible occupational and public health hazards.' *Current Medical Mycology,* 2(1), 36–42. Available at: 10.18869/acadpub.cmm.2.1.36 [Accessed 26 Sep. 2023].
(58) Howell, J. (2008). *The Rising Damp Myth.* London, UK: Nosecone Publications.
(59) Heritage-House. Op cit.
(60) Ward P. Heritage. (n.d). The Warm Dry Home. Available at: www.heritage-survey.org/the-warm-dry-home [Accessed 26 Sep. 2023].
(61) Vanhellemont, J. D. (2021). 'Effectiveness of electromagnetic and electro osmosis methods for the treatment of rising damp.' Abstract from *1st International Conference* on Moisture in Buildings (ICMB21) online. Available at: https://doi.org/10.14293/ICMB210086 [Accessed 26 Sep. 2023].
(62) Citizens Advice. (n.d). Repairs – damp. Available at: www.citizensadvice.org.uk/housing/repairs-in-rented-housing/repairs-common-problems/repairs-damp/ [Accessed 26 Sep. 2023].
(63) Stainton S. and Sandwith H. (2013). The National Trust Manual of Housekeeping: A Practical Guide to the Conservation of Old Houses and their Contents. Supplementary.
(64) May, N., McGilligan, C., and Ucci, M. (2019). *Health and Moisture in Buildings.* [pdf] Available at: https://ukcmb.org/wp-content/uploads/2019/10/health-and-moisture-in-buildings-report-1.pdf [Accessed 26 Sep. 2023].
(65) Two New Projects within the UKCMB. (2016). Available at: https://ukcmb.org/2016/05/20/two-new-projects-within-the-ukcmb/ [Accessed 26 Sep. 2023].
(66) Polygon. (n.d). Polygon – Integrity, Excellence and Empathy. Available at: www.polygongroup.com/en-GB/ [Accessed 26 Sep. 2023].
(67) Leaman, A., and May, N. (n.d). UK Moisture in Buildings Survey. Available at: https://ukcmb.org/2020/05/22/uk-moisture-in-buildings-survey/ [Accessed 26 Feb. 2023].
(68) Leaman, A., and May, N. (2019). 'UK Moisture in Buildings Survey: Conditions, Actions and Outcomes.' *UK Centre for Moisture in Buildings.* Available at: www.usablebuildings.co.uk/UsableBuildings/Unprotected/UKMoistureInBuildingsSurveyMainReport.pdf [Accessed 26 Sep. 2023].
(69) Aktas, Y. D., Altamirano, H., and Ioannou, I. (2018). 'Indoor Mould Testing and Benchmarking: A Public Report.' *UKCMB.* Available at: www.researchgate.net/publication/327338944_Indoor_Mould_Testing_and_Benchmarking_A_Public_Report [Accessed 26 Sep. 2023].
(70) BIB. (n.d) *Publications.* Available at: https://borninbradford.nhs.uk/research/publications/ [Accessed 26 Sep. 2023].
(71) Young, C. (2023). 'Poor quality housing in Bradford branded a 'disgrace'. *Telegraph & Argus.* Available at: www.thetelegraphandargus.co.uk/news/23256987.poor-quality-housing-bradford-branded-disgrace/ [Accessed 26 Sep. 2023].
(72) Bradford Schools Online. (2023). Opportunity for pupils to investigate their classroom air quality. Available at: https://bso.bradford.gov.uk/news/21966-opportunity-for-pupils-to-investigate-their-classroom-air-quality [Accessed 26 Sep. 2023].
(73) SAMHE. (n.d). Meet The Team. Available at: https://samhe.org.uk/about/team [Accessed 26 Sep. 2023].
(74) SAMHE. (n.d). Key Definitions. Available at: https://samhe.org.uk/resources/key-definitions#total-volatile-organic-compounds [Accessed 26 Sep. 2023].

(75) Stanford, M. (2021). 'Bradford ranked in top UK cities for damp and mould issues.' *Telegraph & Argus*. Available at: www.thetelegraphandargus.co.uk/news/19247730.bradford-ranked-top-20-uk-cities-damp-mould-issues/ [Accessed 26 Sep. 2023].

(76) Mulcahy, B. (2013). *Building Health Check Report*. [pdf] Available at: www.wuft.org/news/files/2013/11/SESBuildingHealthCheckReport.pdf [Accessed 26 Sep. 2023].

(77) Hyvonen, S. M. (2020). 'Association of toxic indoor air with multi-organ symptoms in pupils attending a moisture-damaged school in Finland.' *Am J Clin Exp Immunol*. 9(5) pp.101–113.

(78) Madsen, A. M. et al. (2018). 'Concentrations of Staphylococcus species in indoor air as associated with other bacteria, season, relative humidity, air change rate, and S. aureus -positive occupants.' *Environmental Research*, 160, 282–291. Available at: https://doi.org/10.1016/j.envres.2017.10.001 [Accessed 26 Sep. 2023].

(79) Madsen, A. M. et al. (2023). 'Airborne bacterial species in indoor air and association with physical factors'. *UCL Open Environment*, 5. Available at: https://doi.org/10.14324/111.444/ucloe.000056 [Accessed 26 Sep. 2023].

(80) Efthymiopoulos, S. (2021). 'Air sampling and analysis of indoor fungi: a critical review of passive (non-activated) and active (activated) sampling.' *UCL Discovery (University College London)*. Available at: https://doi.org/10.14293/icmb210021 [Accessed 26 Sep. 2023].

(81) Hunter, E.S. et al (2021). 'Effect of patient immunodeficiency's on the diagnostic performance of serological assays to detect Aspergillus-specific antibodies in chronic pulmonary aspergillosis.' *Respiratory Medicine*, 178, 106290. Available at: https://doi.org/10.1016/j.rmed.2020.106290 [Accessed 26 Sep. 2023].

(82) Kosmidis, C., Hashad, R., Mathioudakis, A.G., McCahery, T., Richardson, M.D. and Vestbo, J. (2021). Impact of self-reported environmental mould exposure on COPD outcomes. *Pulmonology*, 29(5), pp. 375–384. Available at: https://doi.org/10.1016/j.pulmoe.2021.05.003 [Accessed 26 Sep. 2023].

(83) University Of Manchester. (2021). During summer, hazardous mould species more abundant in rooms with windows. Available at: www.manchester.ac.uk/discover/news/during-summer-hazardous-mould-species-more-abundant-in-rooms-with-windows/ [Accessed 26 Sep. 2023].

(84) Richardson (2021). Op cit.

(85) Du, C., Li, B., and Yu, W. (2020). 'Indoor mould exposure: Characteristics, influences and corresponding associations with built environment—A review.' *Journal of Building Engineering*, 35,101983. Available at: https://doi.org/10.1016/j.jobe.2020.101983 [Accessed 26 Sep. 2023].

(86) Owen, R. (2022). 'Why are British homes so damp and mouldy?' *Timberwise*. Available at: https://www.timberwise.co.uk/2022/01/why-are-british-homes-so-damp-and-mouldy [Accessed 26 Sep. 2023].

(87) Webb, J.S. et al. (2000). 'Fungal Colonization and Biodeterioration of Plasticized Polyvinyl Chloride.' *Applied and Environmental Microbiology*, 66(8), pp.3194–3200. Available at: https://doi.org/10.1128/aem.66.8.3194-3200.2000 [Accessed 26 Sep. 2023].

(88) Morby, A. (2019). 'Fire experts single out catastrophic Grenfell safety failures.' *Construction Enquirer*. Available at: www.constructionenquirer.com/2018/06/05/fire-experts-single-out-castastrophic-grenfell-safety-failures/ [Accessed 26 Sep. 2023].

(89) Jerónimo, A. (2020). 'Hydraulic lime mortars incorporating micro cork granules with antifungal properties.' *Construction and Building Materials*, 255, pp.119368–119368. Available at: https://doi.org/10.1016/j.conbuildmat.2020.119368 [Accessed 26 Sep. 2023].

(90) Hannah, N., Marincioni, V., Chalabi, Z. and Altamirano-Medina, H. (2021). 'Assessing the performance of Internal Wall Insulation considering transient conditions.' *1st International Conference on Moisture in Buildings 2021 (ICMB21)* Online. 28-29 June. Available at: www.scienceopen.com/hosted-document?doi=10.14293/ICMB210028 [Accessed 26 Sep. 2023].

(91) Brooke-Peat, M. and Glew, D. (2021). 'Unintended Consequences of Internal Wall Insulation; Increased Risk of Mould Growth for Uninsulated Neighbours.' *1st International Conference on Moisture in Buildings (ICMB21)*, 28–29 June, UCL London. Link to Leeds Beckett Repository record: https://eprints.leedsbeckett.ac.uk/id/eprint/7598/ [Accessed 26 Sep. 2023].

(92) Woolley, T. (2022). *Natural Building Techniques*. Crowood Press.

(93) (n.d) 'Retrofit fact file: A short summary of facts and publications relevant to domestic retrofit' [pdf] Available at: https://ppr-website.s3.eu-west-2.amazonaws.com/uploads/2016-URBED-Tyndall-The-Retrofit-factfile-facts-and-publications.pdf [Accessed 26 Sep. 2023].

(94) Shrubsole, C., Macmillan, A. and May, N. (2014). '100 Unintended consequences of policies to improve the energy efficiency of the UK housing stock.' *Indoor and Built Environment*, 23(3), 340–352. Available at: https://doi.org/10.1177/1420326x14524586 [Accessed 26 Sep. 2023].

(95) KATE DE SELINCOURT. (2018). Preston Retrofit Disaster. Available at: www.katedeselincourt.co.uk/preston-retrofit-disaster/#:~:text=Up%20to%20390%20homes%20were,to%20pay%20for%20repairs%20themselves [Accessed 26 Sep. 2023].

(96) Mavrogianni, A. (2013). The unintended consequences of energy efficient retrofit on indoor air pollution and overheating risk in a typical Edwardian mid-terraced house. [pdf] Available at: www.researchgate.net/profile/Clive-Shrubsole/publication/275581121_The_unintended_consequences_of_energy_efficient_retrofit_on_indoor_air_pollution_and_overheating_risk_in_a_typical_Edwardian_mid-terraced_house/links/582d6afb08ae004f74bab3e7/The-unintended-consequences-of-energy-efficient-retrofit-on-indoor-air-pollution-and-overheating-risk-in-a-typical-Edwardian-mid-terraced-house.pdf [Accessed 26 Sep. 2023].

(97) Brambilla, A., Candido, C., and Gocer, O. (2021). 'My home is making me sick! Implications of poor indoor environment quality on mould growth.' *1st International Conference on Moisture in Buildings 2021 (ICMB21).* Online 28–29 June. Available at: www.scienceopen.com/hosted-document?doi=10.14293/ICMB210011 [Accessed 26 Sep. 2023].

(98) CDC. (2021). Basic Facts about Mold and Dampness. Available at: www.cdc.gov/mold/faqs.htm#:~:text=Many%20building%20materials%20provide%20suitable,the%20growth%20of%20some%20molds [Accessed 26 Sep. 2023].

(99) Nilsson, L. O. (2021) 'Vinyl flooring on concrete – a moisture problem only in the Nordic countries?' *1st International Conference on Moisture in Buildings 2021 (ICMB21)* Online 28=29 June. Available at: www.scienceopen.com/hosted-document?doi=10.14293/ICMB210084 [Accessed 26 Sep. 2023].

(100) Chung, I., Ryu, H., Yoon, S-Y and Ha, J. C. (2022). 'Health effects of sodium hypochlorite: review of published case reports'. *Environmental Analysis, Health and Toxicology*, [online] 37(1), e2022006. Available at: https://doi.org/10.5620/eaht.2022006 [Accessed 26 Sep. 2023].

(101) Dengarden. (2023). Top 10 Mold-Resistant Building Materials. Available at: https://dengarden.com/remodeling/Top-Mold-Resistant-Building-Materials [Accessed 26 Sep. 2023].

(102) (2018). CLH Report: Proposal for Harmonised Classification and Labelling: Based on Regulation (EC) No 1272/2008 (CLP Regulation), Annex VI, Part 2 Substance Name: 4,5-dichloro-2-octyl-2H-isothiazol-3-one; [DCOIT] . Available at: https://echa.europa.eu/documents/10162/69ad7753-ba21-6a3f-9866-8e76502f3d65 [Accessed 26 Sep. 2023].

(103) HSE. (n.d). COPR approved biocidal products. Available at: https://www.hse.gov.uk/biocides/copr/approved.htm [Accessed 26 Sep. 2023].

(104) PBC Today. (2017). Render failure is one of the most common claims received. Available at: www.pbctoday.co.uk/news/building-control-news/render-failure-common-claim/30352/ [Accessed 26 Sep. 2023].

(105) Malone, J. (2016). What Causes Render Damage – Cementitious Render Failure (Part Two). Available at: http://buildingdefectanalysis.co.uk/conservation/cementitious-render-failure-part-two/ [Accessed 26 Sep. 2023].

(106) Martel, T. (2021). 'The monitoring of wall moisture in a property retrofitted with Internal Wall Insulation.' *Case Studies in Construction Materials*, 14, Available at: www.sciencedirect.com/science/article/pii/S2214509521000358 [Accessed 26 Sep. 2023].

(107) Owen, A. (2023). Who Applies for Energy Grants? Available at: https://ukerc.ac.uk/news/who-applies-for-energy-grants/#:~:text=Some%20of%20these%20incentives%20are,be%20improved%2C%20or%20boilers%20updated [Accessed 26 Sep. 2023].

(108) Energy Saving Genie. (2023). Grants for Heating and Insulation: Eligible Benefits in ECO4. Available at: https://energysavinggenie.co.uk/energysavinggenie-co-uk-grants-for-heating-and-insulation-eligible-benefits-in-eco4/ [Accessed 26 Sep. 2023].
(109) De Selincourt. (2018). Op cit.
(110) RETROFIT ACADEMY CIC. (n.d). What is the role of a Retrofit Coordinator? Available at: https://retrofitacademy.org/what-is-the-retrofit-coordinator-role/ [Accessed 26 Sep. 2023].
(111) Information provided by Andrea Howe Preston city council energy officer in a power point (unpublished) sent to the author (n.d).
(112) Molloy, E. (n.d). Preston City Council: Private Sector CESP Schemes. Available at: https://apse.org.uk/apse/index.cfm/members-area/advisory-groups/renewables-climate-change/past-presentations/2013/15-may-2013/highways-15-may-prestonpdf/ [Accessed 26 Sep. 2023].
(113) De Selincourt, K. (2018). Disastrous Preston retrofit scheme remains unresolved. Available at: https://passivehouseplus.ie/news/health/disastrous-preston-retrofit-scheme-remains-unresolved [Accessed 26 Sep. 2023].
(114) Baron, O. (2018). House of Commons discuss Fishwick residents left with property damage. Available at: www.blogpreston.co.uk/2018/10/house-of-commons-discuss-fishwick-residents-left-with-property-damage/ [Accessed 26 Sep. 2023].
(115) COMPANY CHECK. (n.d). https://companycheck.co.uk/director/927180923/MR-MARK-GEORGE-BROWNING/companies [Accessed 26 Sep. 2023].
(116) WETHERBY. (n.d). Solid Wall Insulation. Available at: www.wbs-ltd.co.uk/customer-support/external-wall-insulation/ [Accessed 26 Sep. 2023].
(117) NEA. (n.d). Warm and Safe Homes in Fishwick. Available at: www.nea.org.uk/fishwick/ [Accessed 26 Sep. 2023].
(118) K SYSTEMS. (n.d). Available at: https://k.systems/solutions/systems/k-systems-e [Accessed 26 Sep. 2023].
(119) Insulfoam Platinum. (n.d). Premium Insulation Systems. [pdf] Available at: www.insulfoam.com/wp-content/uploads/2014/04/15000-Insulfoam-Platinum-Brochure_REV3-17.pdf [Accessed 26 Sep. 2023].
(120) Civalli (n.d.). Government Insulation Scheme Ruined my Home Government Insulation Scheme Ruined my Home. http://www.civalli.com [Accessed 26 Sep. 2023].
(121) Eshrar, L. Bevan, R., and Woolley, T. (2019). *Thermal Insulation Materials for Building Applications*. London: Ice Publishing.
(122) Housing Executive. (2019). Cavity Wall Insulation Report released for Northern Ireland. Available at: www.nihe.gov.uk/home/news/cavity-wall-insulation-report-released-for-norther [Accessed 26 Sep. 2023].
(123) Advanced DAMP. (n.d). Does Internal Wall Insulation cause Condensation? Available at: https://advanceddamp.co.uk/insulation/does-internal-wall-insulation-cause-condensation [Accessed 26 Sep. 2023].
(124) Latif, E., Bevan, R. and Woolley, T. (2019). *Thermal Insulation Materials for Building Applications* (Chapter 11) Institute of Civil Engineers ICE Publishing.
(125) Osanyintola, O.F. and Simonson, C.J. (2006). 'Moisture buffering capacity of hygroscopic building materials: Experimental facilities and energy impact.' *Energy and Buildings*, 38(10), 1270–1282. Available at: https://doi.org/10.1016/j.enbuild.2006.03.026 [Accessed 26 Sep. 2023].
(126) Latif, E. et al. (2018). 'An experimental investigation into the comparative hygrothermal performance of wall panels incorporating wood fibre, mineral wool and hemp-lime.' *Energy and Buildings*, 165, pp.76–91. Available at: https://doi.org/10.1016/j.enbuild.2018.01.028 [Accessed 26 Sep. 2023].
(127) Latif, E., Lawrence, M., Shea, A. and Walker, P. (2015). 'Moisture buffer potential of experimental wall assemblies incorporating formulated hemp-lime.' *Building and Environment*, 93, pp.199–209. Available at: https://doi.org/10.1016/j.buildenv.2015.07.011 [Accessed 26 Sep. 2023].
(128) Walker. R. and Pavia, S. (2018) 'Monitoring moisture in a historic brick wall following the application of thermal insulation.' *Buildings and Environment*, 133, pp. 176–186.

(129) De Selincourt, K. (2015). 'Risks of Retrofit'. *Green Building.* Available at: www.katedeselincourt.co.uk/wp-content/uploads/2016/01/risks-of-retrofit.pdf [Accessed 26 Sep. 2023].

(130) Kingspan. (2009). Breathability – A White Paper. A study into the impact of breathability on condensation mould growth etc. Second Issue November 2009.

(131) Price, S., Andreou, E., Wilcockson, T., Megagiannis, J. and Lopez, N. (2021). Retrofit Internal Wall Insulation: Guide to Best Practice. [pdf] Available at: https://assets.publishing.service.gov.uk/government/uploads/system/uploads/attachment_data/file/1019707/iwi-guidance.pdf [Accessed 26 Sep. 2023].

(132) Trustmark. (n.d). *PAS 2035: 2019.* Available at: www.trustmark.org.uk/business/information-and-guidance/pas-20352019 [Accessed 26 Sep. 2023].

(133) Trustmark. (n.d). Retrofit your home. Available at: www.trustmark.org.uk/homeowner/information-and-guidance/retrofit-your-home [Accessed 26 Sep. 2023].

(134) Moody, E. (2022) *Updated Guidance for Internal Wall Insulation.* Available at: https://retrofitacademy.org/updated-guidance-for-internal-wall-insulation/ [Accessed 26 Sep. 2023].

(135) QODA. (n.d). Available at: www.qodaconsulting.com/ [Accessed 26 Sep. 2023].

(136) Efficient Building Solutions. (n.d). Available at: www.efficientbuildingsolutions.co.uk/about-us [Accessed 26 Sep. 2023].

(137) QODA. (n.d). Sarah Price| Technical Director. Available at: www.qodaconsulting.com/staff-profiles/sarah-price [Accessed 26 Sep. 2023].

(138) QODA. (n.d). Passivhaus: Genuinely Low Energy Buildings. Available at: www.qodaconsulting.com/resource-list/passivhaus-genuinely-low-energy-buildings [Accessed 26 Sep. 2023].

(139) Morgan, C. (2021). Indoor Air Quality in Airtight homes – A Designer's Guide. Available at: www.seda.uk.net/products/q1urmll4rtkao5add1cyzk3wdjc6hu#:~:text=Airtight%20homes%20can%20overheat%2C%20become,with%20respiratory%20or%20immunity%20problems [Accessed 26 Sep. 2023].

6 Ventilation and personal contaminants

Ventilation is often seen as the main way to deal with indoor pollutants. While good ventilation is essential, it will not necessarily deal with long-term emissions from chemicals in building materials. High levels of solvents following decoration and renovation works can be rapidly diluted through powerful extract or purge ventilation, but this does not necessarily remove all the contaminants. To ensure a healthy indoor environment it is essential to prioritise the use of low-chemical and hazard-free materials, otherwise referred to as source control.

Deniers of building emissions tend to focus on the problem of emissions from cleaning fluids, personal hygiene and air freshener products as the main sources of indoor air pollution and so it is important to consider this aspect of indoor air quality. Chemicals introduced into the home by occupants for personal use can be a serious source of indoor chemical contamination, but these can easily be removed by occupants through ventilation and so they are included in this chapter, as they do not need to be a long-term source of pollution. Even paints, thinners and solvents can be removed rather than being stored inside the home as they may not always be properly sealed. While occupants can take responsibility for these problems it is very difficult for householders to remove building materials that contain chemicals. Before dealing with the issue of ventilation it is necessary to address the issue of personal contaminants.

Air fresheners, scented candles and incense

Many people like to use air fresheners and introduce what they regard as nice smells into their home. These are heavily advertised on television and can be seen as a way to remove or mask unpleasant odours from people, pets and even mould and damp. Introducing perfumed smells into the home does not always eradicate the source of the problem but simply masks it. Some of the chemicals emitted from perfumes contain a range of toxic VOCs and chemicals which can be almost as dangerous as the many chemicals discussed in Chapters 2 and 3. While this is used as a way of denying the dangers of building material emissions, the problem of fragrances is not as serious. However, chemical perfumes if used in large amounts can be absorbed by furnishings and finishes in the home and remain as a long-term emission problem. The building emissions deniers tend to talk a lot about perfumes and other domestic contaminants being dangerous, but they do not appear to call for them to be banned or restricted.

It is possible to introduce natural smells from plants such as lavender and other herbs and some oils, but many of the air fresheners also use this as a marketing device.

DOI: 10.1201/9781003129226-6

Scented candles

All parts of candles from the wax to the wick can be harmful but it is the mass-produced chemical and synthetic-based products that are the most hazardous. Some candle parts can be made from metals which are themselves dangerous when heated. Scented candles should be banned for interior use, but the UK market for candles is worth £1.9 billion and so this is unlikely to happen. But scented candles are perhaps the most serious emitters of hazardous chemicals in many homes. **(1), (2), (3), (4), (5)**

Incense

An extensive literature review by Silva et al lists benzene, toluene, styrene, naphthalene, furfural, furan, isoprene, 2-butenal, phenol, 2-furyl methyl ketone, formaldehyde, acetaldehyde and acrolein in different kinds of incense. The study also indicates that the incense cone type shows a higher probability of being more polluting than incense sticks. Burning candles and incense in churches is said to generate particulate matter (PM) that produces poor indoor air quality and may cause human pulmonary problems as well as being a cancer risk. Incense is also widely used in homes. **(6), (7), (8), (9), (10), (11)**

Air fresheners

There is extensive literature on the dangers of air fresheners, too numerous to list here. There seems to be a high level of ignorance about this among the general public and they are often used in B&Bs, hotels and public buildings as well as homes. Air freshener chemicals are often added to air conditioning units as part of routine servicing and these are not controlled by health and safety legislation. There is a wide range of electrically operated and other kinds but air fresheners should not be confused with air purifiers which are discussed below.

> Air fresheners can impact indoor air quality by adding potentially hazardous pollutants to the air. The use of air fresheners is associated with elevated levels of volatile organic compounds (VOCs), such as formaldehyde, acetaldehyde, benzene, toluene, ethyl benzene, and xylenes, in indoor air. These VOCs are often difficult to smell in the air, but they can irritate the eyes, nose, and throat, as well as cause headaches and nausea. The types and amounts of VOCs emitted depend primarily on the fragrance composition of the air freshener, not on the type of air freshener. It can be difficult or impossible to find out the ingredients in air fresheners, because manufacturers are not required to disclose the complete list of ingredients. **(12), (13)**

Deodorants and sprays

Deodorants and other fragrances applied by aerosol sprays are particularly dangerous and can have serious health effects when applied accidentally in confined spaces or where children are attracted by the smell.

> The parents of a girl (Giorgia from Derby) who died after inhaling aerosol deodorant want clearer product labelling to warn people of the potential dangers. Giorgia Green, who was 14 and was autistic, had a cardiac arrest after spraying the deodorant in her bedroom in 2022. Her parents have since become aware of other young people who accidentally died after inhaling deodorant. **(14)**

As reported by the BBC, the Office for National Statistics (ONS) note that "deodorant" was mentioned on 11 death certificates between 2001 and 2020 but deaths could be higher than this as specific substances are not always mentioned on death certificates and can be blamed on aerosols rather than deodorants. Giorgia Green's death certificate referred to "inhalation of aerosol" rather than "deodorant". Butane – the main ingredient of Giorgia's deodorant – was recorded as having been involved in 324 deaths between 2001 and 2020. The Royal Society for the Prevention of Accidents (ROSPA) has warned about the dangers of inhaling large quantities of aerosols and there have been other deaths such as that of 12-year-old Daniel Hurley, also from Derbyshire, who died in 2008 of a deodorant-related incident in a bathroom. Another death, Jack Waple in 2019, sprayed deodorant when he was anxious, as it smelled like his mother. **(15)**

The British Aerosol Manufacturers' Association (BAMA) report that aerosols are made to the highest safety standards and are labelled with very clear warnings and usage instructions and recommend that anyone using an aerosol does so in accordance with the manufacturer's instructions, but there are calls for the warnings to be improved. **(16)** Scented sprays are commonly used in buildings and plugged in fresheners are widely used.

Perfumes

A simple way to get exclusive use of a lift or a toilet in an office building is to arrive reeking of aftershave or perfume. Many more people are sensitive and react badly to such strong odours today, but limiting this relies on social and cultural attitudes which have also changed in recent years. The fragrance market in the UK had a sales value in 2022 of £1.8 billion, and fragrances are added to a wide range of products not just perfume and aftershave but even refuse bags.

> The main ingredients of fragrances are dimethyl octanol, eucalyptol, dimethyl heptenal, alpha-citroneline, methoxy benzaldehyde, limonene, benzylacetone and decanal, terpenes, limonene, hexylcinnamol and linalool, Toluene, benzene , styrene methyl ethyl ketone and butyl acetate Diethyl phthalate (DEP), di-n-butyl phthalate, di (2-ethylhexyl) (DEHP) and butylbenzyl phthalate (BBzP) P-t-butyl-α-methyl hydrocinnamic aldehyde, eugenol, geraniol, hydroxycitronellal and ethanol, diethyl phthalate Terpenes. **(17)**

Some refer to perfumes as poisons:

> You might think that finding out what's in your perfume would be as easy as reading the ingredient label. But because of laws that protect fragrance manufacturers from sharing "trade secrets," almost every perfume sold commercially is crammed with chemicals that aren't listed individually on the product packaging. **(18)**

Fragrances included in products discussed above and bought over the counter in city centre shops can be very dangerous to health. Particularly high concentrations of emissions from perfumes will be experienced while walking through duty-free shops in airports though it is possible in some airports like Gatwick North terminal to bypass the choking emission concentrations using the accessibility channel. The author recent discovered there is a bypass to the perfume alley at Stansted airport, kindly pointed out by a member of the airport staff, by following the red fast track signs.

The Canadian Centre for Occupational Health and Safety has introduced scent-free policies for the workplace under Human Rights legislation:

Table 6.1 List of symptoms associated with scented products

headaches
dizziness, light-headedness
nausea
fatigue
weakness
insomnia
numbness
upper respiratory symptoms
shortness of breath
skin irritation
malaise
confusion
Exposure to cosmetic chemicals can be very dangerous to pregnant women and staff exposed on a daily basis in retail places of work. Pthalates can be found in high levels among workers who work with cosmetics all day. **(24)**

> They list a range of health problems including headaches, dizziness, light-headedness, nausea, fatigue, weakness, insomnia, numbness, upper respiratory symptoms, shortness of breath, skin irritation, malaise, confusion, difficulty with concentration.
>
> Ingredients or chemicals used to produce scents are present in a very large range of products, including shampoo and conditioners, hairsprays, deodorants, colognes and aftershaves, fragrances and perfumes, lotions and creams, potpourri, industrial and household chemicals, soaps, detergents, fabric softeners, cosmetics, air fresheners and deodorizers, oils, candles and diapers (Nappies). **(19)**

Singal et al conduct a useful literature review with critical comments on Steinemann, etc., and discuss the contradictions between chemicals emitted when peeling an orange with the chemicals emitted from fragrances. **(20), (21), (22)**

Radis-Baptista warns that fragrance chemicals interfere with neuro endocrine-immune axis promoting cancer and developmental problems.

> The increased concentration of fragrances and fragranced-associated VOCs in the indoor air may cause adverse cutaneous, respiratory, and systemic effects such as headaches, asthma attacks, breathing difficulties, cardiovascular and neurological problems, mucosal irritation, and contact dermatitis, as well as distresses .... **(23)**

Exposure to scented products can affect a person's health, including the symptoms listed in Table 6.1. **(24)**

Leftover paints and solvents

Many houses contain a host of old tins of paint and solvents, often not properly sealed, not just in garages and sheds but in kitchen and broom cupboards. These can emit VOCs and other chemicals and should be removed once used. The Royal Society of Chemistry estimates households "stash" over 50 million litres of paints in their homes. **(25)**

Cleaning fluids

> Many cleaning supplies or household products can irritate the eyes or throat, or cause headaches and other health problems. Some products release dangerous chemicals, including volatile organic compounds (VOCs). VOCs are chemicals that vaporize at room temperature. Even natural fragrances such as citrus can react to produce dangerous pollutants indoors. VOCs and other chemicals released when using cleaning supplies contribute to chronic respiratory problems, allergic reactions and headaches. **(26)**

It is possible to select low odour and relatively non-toxic cleaning products but many people feel they are not as effective and some ecoproducts are more expensive. The UK Health and Safety Executive lists hazardous products including cleaning agents but they are not able to police what is sold over the counter in supermarkets. **(27), (28)**

Clearer guidance on safe cleaning materials is provided by Government agencies in the USA, whereas in the UK guidance is generally vague and ineffective.

Hospitals and care homes can have the most aggressive cleaning regimes due to concerns about infection, but public health bodies do not issue clear guidance on what cleaning materials should or should not be used, despite warning about the dangers of some products. **(29)**

Pollution and air conditioning

Good ventilation is essential for good indoor air quality. Unfortunately, not everyone is surrounded by clean fresh air. Many urban areas suffer from traffic, industrial and aircraft pollution and even in rural areas there are worryingly high levels of ammonia pollution from the poorly controlled spreading of manures, slurries and fertilisers. Pesticides are also spread on crops near houses are even more dangerous. This does not mean that we all have to seal up our houses and introduce major filtration systems to provide clean air to breathe. Opening windows to purge ventilate houses is still an important and useful measure. Filtering air is not always the best solution unless filters are of good quality and regularly cleaned or replaced.

As temperatures rise with global warming, with excessive temperatures reached during the summer of 2023, more and more people are installing air conditioning units. However, air conditioning units are not a means of ventilation, they do not bring in outside air but recirculate existing indoor air with obvious health problems. This air which is sucked into a unit is cooled using a great deal of electrical energy. There are some kinds of AC units that bring in outside air but these are less common. The warm air sucked into the unit and any associated moisture is then expelled to the outside and may contribute to bad external air quality around buildings.

> … not all AC systems are able to introduce fresh air to the mix as they are only intended to recirculate and cool indoor air. **(30)**

From time to time there are outbreaks of Legionnaires disease as a result of people breathing in the Legionella bacteria, usually from microscopic water droplets. These can come from showers, taps and ventilation systems (mostly in larger buildings). Hot tubs, swimming pools and even birthing pools can be a source. Water hygiene is essential to reduce risks. An accommodation barge intended for asylum seekers had to be evacuated when Legionella was identified. **(31)**

Ventilation and indoor air quality

Natural ventilation can be all that is required for good indoor fresh air but modern buildings that are sealed up with plastic materials can develop high levels of humidity which are bad for health and will encourage damp and mould. Some provision of mechanical ventilation will therefore be necessary to reduce these risks. It is not possible to provide comprehensive guide to ventilation in this book, despite its importance to IAQ, but there are a few important issues to note. For many experts indoor air quality (IAQ) is largely about ventilation with the importance of emissions from building materials underplayed or even ignored.

Baker and Steemers, in their much quoted 220-page book, *Healthy Homes,* devote only 6 pages to indoor air emissions and largely in terms of ventilation. Building materials get a very brief mention on page 127, with far more space devoted to occupants, animals, smoking, cooking, vacuum cleaning and mould. They also make a strange criticism about natural materials "as they may have been treated with preservatives, fungicides, insecticides or polishes," but they don't make any criticisms of the wide range of chemical and hazardous materials that most buildings are built from. Their book does provide a useful but short guide on mechanical ventilation but provides a very different perspective on healthy homes to this book by ignoring building materials and emissions. **(32)**

The Chartered Institute of Building Engineers (CIBSE) is the main professional body providing training and guidance on ventilation. Their publication, "Indoor Air Quality and Ventilation," was published in 2011. It gives a general outline of indoor air pollutants and a good outline of ventilation systems as summarised in Table 6.2.

Good indoor air quality is seen by CIBSE largely in terms of ventilation but they do refer to the selection of materials to minimise pollutant emissions. **(33)**

The American Society of Heating, Refrigerating and Air-Conditioning Engineers (ASHRAE) in the USA provides a more comprehensive standard for ventilation and indoor air quality. Its standard 62.2 for residential buildings was updated in 2022 and deals with mechanical exhaust and other forms of ventilation. It provides a useful but highly technical guide to infiltration and ventilation rates. **(34)**

While ASHRAE, like CIBSE, sees IAQ mainly in terms of ventilation, a recent ASHRAE statement on indoor air quality does include a general discussion of source control:

> Use building materials, furnishings, appliances, and consumer products with low contaminant emissions;

Table 6.2 Outline of indoor pollutants and ventilation systems

Natural ventilation
Mechanical ventilation
Mixed mode ventilation
Required ventilation flowrates for good IAQ
Fresh air prescribed flow rates
Flows required for thermal comfort (heating)
Flows required for thermal comfort (cooling)
Flows between spaces and pressurisation
Pollutant removal
Measuring ventilation and IAQ
Measurement of flowrates

Minimize indoor contaminant sources caused by occupant activities;

Remove outdoor contaminants via filtration and air cleaning before they enter a building; and

Design, operate, and maintain building enclosures, HVAC systems, and plumbing systems to reduce the likelihood of moisture problems and/or quickly mitigate them when they happen. **(35)**

Mechanical ventilation

The mechanical ventilation industry is very active in calling for better indoor air quality, but in terms of promoting mechanical ventilation as the solution. They have published a series of guides which are very useful in terms of understanding the need for good ventilation. However, their guidance on existing homes includes recommending Positive Input Ventilation (PIV).

> A properly designed and installed PIV system will effectively mix the indoor air with external air throughout a dwelling and gently change the air in a home on a controlled and continuous basis. The continuously diluted internal air has a resulting lower moisture content and therefore dew point. **(36)**

Many PIV installations simply involve a fan in the ceiling of the house, blowing air from the loft into the house. There are great dangers in this approach in that the air will be contaminated with particulates from (often dirty) glass fibre insulation in the loft. Promoting PIV is not going to improve indoor air quality in some circumstances and may transfer greater levels of moisture from the roof into the house.

While good ventilation is essential for human health, mechanical ventilation is not necessarily the panacea for good indoor air quality. Ventilation can be natural, active or passive and mechanical as well, but ventilation needs can vary depending on the nature of the materials and construction used in buildings. Buildings sealed with air-tight non-vapour permeable materials may require a higher level of ventilation to try and reduce moisture content and noxious emissions. Ultra-low energy buildings such as those advocated by the passivhaus movement (see below) are dependent on mechanical ventilation systems which involve extract fans, heaters and duct work. These are referred to as Mechanical Ventilation and Heat Recovery systems (MVHR).

An important study, commissioned by the UK Government, was carried out to evaluate whether the ventilation provisions in the UK building regulations would provide satisfactory indoor air quality in new homes. The study reviewed 80 recently built homes, 55 of which were naturally ventilated and 25 had decentralised mechanical extract systems. The study found poor indoor air quality in many of the homes, particularly in bedrooms, and a failure to meet regulatory requirements. Residents complained about the noise of extract systems and noise coming from outside. **(37)**

Despite the existence of regulations for ventilation, which the AECOM study implies might not be adequate anyway, the other problem is that ventilation performance is not tested when new houses are built. There is a requirement to test for air tightness but not for ventilation. The introduction of increased air tightness resulted from aims for greater energy efficiency without much consideration being given to unintended consequences such as bad indoor air quality. Despite increasing emphasis on the importance of air tightness, so that homes are more fuel-efficient, the resulting lack of clean, fresh air could create serious health risks, according to *Marrs in the Architects Journal.*

Little has been done since 2011 to address this problem despite further reviews of building regulations, which have continued to focus on greater energy efficiency. **(38), (39)**

Recent consultation documents on changes to the building regulations raise concerns about indoor air quality and suggest the possibility of installing CO2 monitors in buildings. This is a measure being introduced in Scotland, though the CO2 monitors will not have alarms and depend on the householder checking them to make sure their CO2 limits are not too high. The installation of CO2 monitors is also being considered in English building regulations.

The risk of indoor air problems due to increased air tightness does appear to have been recognised in some discussions of future changes to the building regulations, and a suggestion for some alterations to improve assessment has been put forward, based on concentrations of individual compounds using *Public Health England's Indoor Air Quality Guidelines for Selected Volatile Organic Compounds (VOCs) in the UK,* but as discussed in a previous chapter, there is no rationale for only selecting a handful of VOCs. Reference to the Public Health England (PHE)-selected VOCs was also included in draft energy regulations in Northern Ireland. **(40), (41)**

An organisation, called the Zero Carbon Hub, set up an indoor air quality task group in 2013 funded by the National Housing Building Council. This group found very few case study examples that met airtightness standards or good indoor air quality. They published a report on Mechanical Ventilation with Heat Recovery (MVHR). Their study provided the mistaken basis for assuming that good indoor air quality is measured through ventilation but it did include a reference to "toxic pollutants" without any explanation as to what this meant. **(42)**

The Good Homes Alliance (GHA) provides guidance on ventilation and indoor air quality but only mentions emissions from building products once in 90 pages (2011). The GHA also offers an online tutorial series on delivering health homes with Tim Sharpe, which concentrates on ventilation. **(43)**

Extract ventilation

Many houses will have inadequate extract ventilation. While purge ventilation can be a valuable way to reduce smells and bad air it is not always appropriate. Small extract fans in bathrooms that only come on when the light is switched on are generally inadequate and yet have been considered acceptable by the building regulations for many years. More expensive extracts include humidistats which will boost the extraction levels as moisture levels rise. These can be left in trickle mode throughout the day and night though some people do not like the noise these extracts make when they respond to increased moisture levels. The ventilation industry has come on by leaps and bounds in recent years and demand control ventilation. Humidity-sensitive extracts can come in many shapes and sizes as well as air inlets that are draught free. **(44), (45), (46)**

Cooker extracts are essential to maintaining good air quality, particularly as so much emphasis has been placed on suggesting that emissions from cooking are dangerous. Unfortunately, many homeowners and landlords install *recirculating* cooker hoods. These look like extract hoods but simply provide a fan which sends the air through a filter and back into the kitchen. Recirculating hoods do not deal with steam and moisture so tend to increase humidity levels in kitchens. Using a well-maintained cooker extract hood (where the filters are regularly cleaned or changed) will ensure that most emissions from cooking will be extracted to the outside air.

It was difficult to find research evaluating recirculating hoods, although one back in 1986 suggested that Nox and PM2.5 could be reduced by 30%. A fresh carbon filter removed 60% of NO2 but thus dropped to 20% after a few weeks. **(47), (48)**

A general guide to cooking emissions by Surrey University includes the statement that

> using extractions fans and keeping doors and windows open can reduce the average in-kitchen particulate matter exposure by about 2-times … . **(49)**

Does ventilation reduce off-gassing?

Some academic studies claim that there are chemicals in buildings (with high emissions in the first two years) that can be reduced to an acceptable level through the use of mechanical ventilation. The suggestion is that chemical emissions will decline to a level that is not a risk to health after the initial 2 years. It is even suggested by some "experts" that buildings should be left unoccupied for a period from a few days to 6 months and that emissions should be "baked off" by overheating the building to 30 degrees so that volatile organic compounds, for instance, will be given off at higher rates at higher temperatures. **(50), (51)**

Studies at Nottingham University have tried to identify the basis for harm caused by airborne contaminants in buildings by establishing an acceptable level related to ventilation rates. Leaving aside the question of whether the philosophical concept of an *acceptable* level of harm is acceptable, this research does not consider source control…removing the harmful contaminants in the first place, despite some excellent epidemiological and risk assessment work on the health effects. Morantes et al identify contaminants such as Acrolein 0.045, Benzene 0.0012, Formaldehyde 7.6, Mold 0.013, Nitrogen dioxide 2.7 Ozone 0.072, PM10 290, PM2.5 100, Radon 0.3 and Sulphur dioxide, but they seem to assume that the presence of contaminants in buildings is inevitable, but can be dealt with through ventilation. **(52)**

In numerous other scientific papers on ventilation it has not been possible to find any suggestion that chemicals emitted from products should not be used. Presumably it is assumed that hazardous chemicals are a normal and acceptable part of building construction and the problem is how to reduce their impacts. All of these methods impose a higher energy burden and cost on buildings which is the result of much increased mechanical ventilation and leaving buildings empty for half a year. Not surprisingly this is not often done!

There are some fundamental questions hidden away here. Firstly, the assumption that emissions will decay to acceptable levels over time. The Building Emissions Deniers (BEDs) rely heavily on this assumption, though as has been discussed, the science of decay is underdeveloped. Most of the scientific studies on off-gassing rely on TVOC analysis and rarely distinguish between many hundreds of chemicals and toxic substances. Also, there has been very little research into the reaction of different people to off-gassing chemicals. Some people may be hyper-sensitive to much lower levels of chemicals than others. Off-gassing lingering in buildings for many years may continue to have negative health effects on some people who are more sensitive whereas others may not be affected.

As discussed above, the ventilation industry continues to promote the idea that off-gassing and chemicals can be removed through ventilation. There is no doubt that ventilation, particularly forced ventilation, can reduce or dilute some concentrations of chemicals in the air, but it doesn't remove the source. Some chemicals used in buildings are scientifically described as persistent, referred to as forever chemicals, which are now well known, and the absorption of chemicals into the body through different means, including thyroid and lungs, may continue over many years despite good ventilation.

Exposure limit values

ASHRAE, referred to above, in addressing bad indoor air quality has established exposure limits:

> The ASHRAE Standing Standard Project Committee on Ventilation and Acceptable Indoor Air Quality in Residential Buildings (62.2) has embarked on a ground-breaking endeavour—(to) put forth a potential addendum to the standard that would add a harm-based indoor air quality procedure as an alternative compliance method.
>
> An airborne contaminant is a substance not normally present in the air. Some contaminants are emitted internally by building materials and furnishings and by occupants and their activities. Other contaminants are brought inside by ventilation and infiltration. A person is exposed to airborne contaminants when they occupy the same space as contaminated air, and their exposure is a function of the concentration and the time spent in the space. The dose received is a function of breathing rate, breathe volume and uptake by the lungs. These factors are affected by metabolic rate and physiology that may be a function of age and sex.
>
> Many standards further increase ventilation rates to minimize contaminant exposures and, therefore, protect occupant health. **(53)**

The concept of using DALYs (Disability Adjusted Life Years) has been adopted by some scientists such as Morantes et al to assess the health impact of bad indoor air quality in an effort to establish limits. Based on an extensive literature review, their aim is to develop what they describe as *health-centred IAQ metrics.* Based on a population of 100,000:

> Total pooled DALYs were estimated per 100,000 population with corresponding uncertainty intervals; Estimated population-averaged annual cost, in units of DALYs lost, of chronic air contaminant inhalation in dwellings, indicate that the contaminants with the highest median pooled DALY loss estimates are PM10 [1.9 103 (95% CI 4.4 102–8.7 103)] and PM2.5 [1.5 103 (95% CI 5.3 102–4.4 103)]. PM coarse, formaldehyde, NO2, radon and ozone have medians among 102–101. Acrolein and SO2 are within 100. Mould-related bioaerosols could still be of interest having >0.5 DALYs per 100,000 exposed population. **(54)**

Such work, in attempting to develop a way of measuring indoor air quality risks, is important but is being carried out within a ventilation context and does not appear to link up with scientists measuring source emissions.

Ventilation systems can introduce fresh air from outside buildings, but may also be introducing external pollutants at the same time, which are identified in the DALY metric. While filters (that have to be frequently changed) can reduce some contaminants such as particulates, they may not remove more noxious chemicals. Some ventilation systems may be recirculating indoor air complete with existing indoor contaminants. While some of these may be mitigated by filters and diluted by fresh air from outside, emissions from building materials may still remain in the building. This is a particular problem affecting air conditioning systems, and these are likely to become more prevalent with hotter summer temperatures even in temperate countries.

Purge ventilation and natural ventilation

> Another approach to lowering the concentrations of indoor air pollutants in your home is to increase the amount of outdoor air coming indoors. Most home heating and cooling systems, including forced air heating systems, do not mechanically bring fresh air into the house.

> Opening windows and doors, operating window or attic fans, when the weather permits, or running a window air conditioner with the vent control open increases the outdoor ventilation rate. Local bathroom or kitchen fans that exhaust outdoors remove contaminants directly from the room where the fan is located and also increase the outdoor air ventilation rate. For most indoor air quality problems in the home, source control is the most effective solution. **(55)**
>
> A common misconception is that dilution with fresh ventilation air is the only way of removing harmful contaminants from the occupied space. It is more effective to control the source of pollutant, if possible, especially as outdoor air may itself be contaminated. Preventing indoor air quality problems by source control is generally less expensive than identifying and solving them after they occur. **(56)**

Using natural ventilation by opening windows or keeping trickle vents open is the best way to ensure better air quality except in a few places where houses are near busy roads with bad traffic emissions. Purge ventilation involves throwing windows open wide each morning but people in the UK seem not so keen on doing this. Opening windows and trickle vents is assumed to be wasting heat. Trickle vents in windows are often perceived to cause drafts, but when replacement windows are fitted, particularly pvc, no trickle vents are included anyway. Trickle vents in roof lights are particularly effective as they are less likely to produce drafts but not all buildings have roof lights. Many houses have different forms of extract fans which are expelling bad air to the outside, but they rely on fresh air entering from somewhere. Fans which come on when the bathroom light is switched on are not very effective and more sophisticated extracts which use humidistats are much better as they come on automatically when moisture levels increase.

Air purifiers and plants

As people become more aware of air pollution, some are buying so-called air purifiers or air cleaners which contain HEPA filters. These devices are currently unregulated, and there are concerns that they are not very effective, while using a lot of energy. HEPA means "high efficiency particulate air" and are only designed to deal with particulates and not other indoor contaminants.

There is a belief that plants purify air but the scientific consensus is that this is a myth: www.gardenmyths.com/garden-myth-born-plants-dont-purify-air/

Indoor plants are attractive and create a feeling of well-being, but they don't deal with pollutants sadly.

Air purifiers are being installed in schools as the UK Government has encouraged their use and some are being monitored in research projects. Air purifiers simply recirculate internal air and depending on the quality of filters can reduce dust, pollen and possibly bacteria. However, they do not deal with chemical pollutants. Given the increased awareness of indoor air problems among the general public, the air purifier companies are cashing in. Very few manufacturers publish data on the performance of their devices but one company, Medicair, supplies what they claim to be medical-grade air purifiers and make an extensive range independent assessment of their products available. Their devices include HEPA and carbon filters and UV lights. They claim that this can help to clean up mould in the home. The independent assessment seems to have been carried out in the USA and Dublin and by various agencies. **(57), (58)**

More on mechanical and natural ventilation

The mechanical ventilation industry is very active in promoting its products but in an extensive review of the literature on ventilation, Wargocki says that natural ventilation can be just as effective:

> ...it seems likely that health risks may occur when ventilation rates are below 0.4 air changes per hour in existing homes. No data were found indicating that buildings having dedicated natural ventilation systems perform less well than the dwellings in which mechanical ventilation systems are installed. **(59)**

While good ventilation can dilute the concentrations of indoor air pollutants, some chemicals are not dealt with as easily.

> New research by the National Center for Healthy Housing (NCHH) and Enterprise Community Partners has found that mechanical ventilation systems significantly reduce the levels of particulate matter, carbon dioxide and monoxide, and formaldehyde from indoor air, reducing the risk of respiratory and cardiovascular illness. However, even in homes with ventilation systems, researchers did not observe significant changes in levels of nitrogen dioxide. **(60)**

Overheating and mechanical ventilation

There is a growing problem of overheating with modern energy-efficient forms of construction which use lightweight materials. There are suggestions that passivhaus projects are more prone to overheating risks and MVHR systems can often be out of balance and thus not as effective. Also, it is often difficult to maintain filters as the equipment is not accessible. There is a growing body of literature raising questions about the limitations of mechanical ventilation and passivhaus. **(61), (62), (63), (64), (65)**

A major study of low energy houses, many built to passivhaus standards, suggests that MVHR leads to better ventilation than natural venting, but not without significant risks that systems may not function effectively:

> Whilst it may be a reasonable expectation that a house with a constantly running mechanical ventilation system can deliver better air change rates – and the performance data suggests that this stabilises internal environmental conditions – a significant question remains as to what conditions occur when the system has suboptimal air flow or is not in use. It is apparent that within the general trends there are some outliers in which environmental conditions in houses with MVHR are very poor – this may be due to a poorly performing system, but the likelihood is that the system is disabled.

> Effects on health – whilst relationships between health and ventilation are well established, robust data on actual effects on health of varying ventilation strategies are limited. Further study is needed on effects on occupant health, comparison of indoor and external pollutants, off gassing from ducts, effects of leaks within systems and pollutant source control. **(66), (67), (68)**

In Scotland, as a result of research by Tim Sharpe and others and the Health Effects of Modern Air Tight construction (HEMAC) group, which discovered high levels of CO2 in bedrooms when people were sleeping at night without ventilation, the Scottish Government is requiring CO2 monitors to be installed in bedrooms (though without alarms):

- Carbon dioxide levels are often used as an indicator of indoor air quality. Measuring CO2 levels is easier and cheaper and monitoring can be done over an extended period.

- Commonly over 1000 ppm at night. **(70)**

The Sharpe Meta study of MVHR systems, referred to above, mentions possible effects from off-gassing from ducts. This is a significant problem which is largely glossed over in literature about ventilation. While MVHR systems claim to provide passive heat recovery this is rarely as effective as claimed, so that systems include a heater to warm the air before it enters the dwelling. One of the reasons that passivhaus projects are not as low energy as claimed is that householders use this heater to heat their house. A range of materials are used in the ductwork including galvanised metal, aluminium polyethylene and expanded polyethylene (which was a major contributor to the Grenfell fire), insulated foil ducting using glass fibre, polyurethane (also flammable), polyester, pvc and expanded polypropylene. Heating the air entering the duct system could increase the rate of emissions from the duct materials, which are then blown into the house.

> Higher indoor temperatures can also lead to changes in IAQ through an increase in concentrations of indoor sourced pollutants, specifically VOCs and a balance needs to be struck between airtightness to prevent ventilation heat loss for greenhouse gas reduction policies and the need for a healthy air change rate. **(71)**

Health problems from duct materials?

There is little doubt that over time dust and microparticles can be accumulated in ducts, with the vibration of fans and heaters and problems with static. Air becomes positively ionised in ducts, which is not good for health, and work has been done to introduce negative ionisation to reduce the impact of particulates. Air ionisers are used by some people to improve indoor air. While their effectiveness has not been proven, the negative effect of positive ions is said to include unpleasantness, acute respiratory irritation and joint symptoms. **(72), (73)**

This is an area that could benefit from more research as there seems to be little data on possible emissions from ducts. It would seem likely that galvanised steel would have fewer emissions issues, but this is not the view of the UK Green Building Council (UKGBC).

The UKGBC promotes a product called Ecoduct on its website that refers to negative problems with many duct materials. **(74)**

"Ecoduct" is said to be a non-metallic pre-insulated air duct and examination of its product data sheet says it is made from "Phenolic". Ecoduct criticises the performance of galvanised steel ducting on its website and suggests that lighter weight phenolic foam is better, though it does imply that the product is flammable on its product data sheet. The company, when contacted, provided a health and safety data sheet from Kingspan Koolduct which turns out to be the same product. There is little information in the Kingspan COSSHH sheet other than an admission that it is a phenolic material, and no detail as to the chemical constituents. Instead, they state that

> There are no applicable Globally Harmonised System (GHS) hazard statements for the substances contained in these products. Accordingly, there are no known hazards associated with the normal handling or use of these insulation products.

The data sheet states that the Reach Regulation does not require a safety data sheet to be provided for the product. Another document from Ecoduct says that the product surface contains halogenated flame retardants, but the Kingspan sheet says there is no data available on endocrine-disrupting properties! Not very reassuring! **(73), (74), (75), (76)**

UKGBC were unwilling to comment on the nature of the Ecoduct products as they said that they do not commercially endorse these products in their Solutions library. (There is a disclaimer on their Solutions page, but it does not appear on the Ecoduct page.) Their media office say that items on their Solutions page are shared primarily for "inspiration and awareness." **(77)**

Risks from possible chemical emissions from ducts and the possible health drawbacks of positive ionisation may be a strong reason for avoiding MVHR and other ducted ventilation systems.

Building regulations, ventilation and indoor air quality

The UK building regulations (closely related to the Irish regulations) lay down requirements for mould and indoor air pollutants (see Table 6.3). These are included in the approved document (Technical booklet K in Northern Ireland and F in England). The wording in the guidance suggests that mould and internal air pollution is linked to ventilation, without any mention of source control or emissions from materials. The inclusion of the term "no visible mould" ignores the established scientific evidence that mould need not always be visible.

> A3 (NI) The performance criterion for moisture is that there should be no visible mould on external walls in a properly heated dwelling with typical moisture generation.

> F1 (*1*) The aim of requirement F1(1) is to protect the health of occupants of the building by providing adequate ventilation. Without adequate ventilation, mould and internal air pollution might become hazardous to health.

Section A5 in the Northern Ireland regulations state that mould growth can occur whether the dwelling is occupied or unoccupied, so the performance criteria for moisture (as set out in Table A2) should be met at all times, regardless of occupancy. **(78)**

Guidance on indoor air emissions varies but appear to be related to WHO guidelines and the "Indoor Air Quality Guidelines for Selected Volatile Organic Compounds" report by Shrubsole et al, which points out that there are no indoor emission guidelines for individual VOCs in the UK. **(79)**

The regulations set out ventilation standards which are primarily aimed at reducing moisture pollution but also imply that this will also remove VOC pollution. The Northern Irish guidance admits that this may not be adequate to address pollutants from occasional, occupant-controlled

Table 6.3 Guidance values in the UK

Carbon monoxide (should not exceed) 100 mg/m3 (90 ppm) – 15 minute averaging time (DOH, 2004); 30 mg/m3 (25 ppm) – 1 hour averaging time (DOH, 2004); or 10 mg/m3 (10 ppm) – 8 hours averaging time (DOH, 2004
nitrogen dioxide (NO2) should not exceed 200 in England but 288 µg/m3 in NI 1 hour average 40 µg/m3 (20 ppb) – long-term (1 year) average
Total Volatile Organic Compound (TVOC) levels should not exceed 300 µg/m3 averaged over 8 hours

activities such as painting, smoking, cleaning or other highly polluting activities. They say they these are not ready for inclusion in this guidance, and indeed they may be better controlled at source (e.g., by avoidance, isolation or the use of fewer emitting products). This is the only place where source control is mentioned.

There is a great deal more work required both in the area of reducing pollutants that can be brought into buildings and effective ventilation standards. Greater use of natural ventilation in along with construction from low emission or natural materials may create healthier homes and buildings as discussed in Chapter 8.

References

(1) IQ Air The Hidden Dangers of Scented Candles. Available at: www.iqair.com/newsroom/hidden-dangers-scented-candles [Accessed 22 Sep. 23].

(2) Massoudi, R. and Hamidi, A. (2017). Some candles emit hazardous materials for human health and are indoor air pollutants. *International Journal of Tropical Disease and Health*, 24(2), 1–10. Doi: 10.9734/IJTDH/2017/34965

(3) Rasmussen, B., Wang, K., Karstoft, J. G., Skov, S. N., Køcks, M., Andersen, C. (2021). Emissions of ultrafine particles from five types of candles during steady burn conditions. *Indoor Air*. DOI: 10.1111/ina.12800.

(4) Ngriagu, J. et al. (2000). Emissions of lead and zinc from candles with metal-core wicks. Science of the Total Environment. https://pubmed.ncbi.nlm.nih.gov/10811249/ [Accessed 22 Sep. 23].

(5) Ahn, J-H. et al. (2015). Characterization of hazardous and odorous volatiles emitted from scented candles before lighting and when lit. *Journal of Hazardous Material.* Epub 2014 Dec 31.

(6) Silva, G. V. et al. (2021). 'Indoor air quality: Assessment of dangerous substances in incense products.' *International Journal of Environmental Research and Public Health*, 18(15), 8086.

(7) Chuang, H-C. (2011). 'Combustion particles emitted during church services: Implications for human respiratory health.' *Environment International*, 40, 137–142. https://pubmed.ncbi.nlm.nih.gov/21831441/ [Accessed 22 Sep. 23].

(8) Cancer Council. (n.d). Is Burning Incense Linked to Respiratory Cancers? Available at: www.cancer.org.au/iheard/is-burning-incense-linked-to-respiratory-cancers [Accessed 22 Sep. 23].

(9) Harder, B. (2006). Holy Smoke. Available at: www.sciencenews.org/article/holy-smoke-burning-incense-candles-pollute-air-churches [Accessed 22 Sep. 23].

(10) Cancer Warnings Over Incense Become Burning Health Issues (2003). Available at: www.heraldscotland.com/news/12534266.cancer-warnings-over-incense-become-burning-health-issue/ [Accessed 22 Sep. 23].

(11) White A. (2018). Is Burning Incense Bad for Your Health? Available at: www.healthline.com/health/is-incense-bad-for-you [Accessed 22 Sep. 23].

(12) Air Fresheners and Indoor Air Quality. Available at: https://ehs.umass.edu/air-fresheners-and-indoor-air-quality [Accessed 22 Sep. 23].

(13) Ajasa, A. (2023). If You Can Smell Your Air Freshener You Might Have a Problem. Available at: www.washingtonpost.com/weather/2023/02/22/air-freshener-indoor-air-quality/ [Accessed 22 Sep. 23].

(14) Lowbridge, C. (January 28, 2023). *BBC News*. Available at: www.bbc.co.uk/news/uk-england-derbyshire-62078939 [Accessed 22 Sep. 23].

(15) Norfolk boy died after inhaling deodorant 'that smelt like his mother' (2019). Available at: www.bbc.co.uk/news/uk-england-norfolk-50087690 [Accessed 22 Sep. 23].

(16) The British Aerosol Manufacturers Association. (n.d). Available at: www.bama.co.uk [Accessed 22 Sep. 23].

(17) Kazemi, Z. et al. (2022). Evaluation of pollutants in perfumes, colognes and health effects on the consumer: A systematic review. *Journal of Environmental Health Science and Engineering*, 20(1), 589–598.

(18) Watson, K. (2019). How to Know If Your Perfume is Poisoning You. Available at: www.healthline.com/health/perfume-poisoning [Accessed 22 Sep. 23].

(19) Scent Free Policy for the Workplace. (n.d). Available at: www.ccohs.ca/oshanswers/hsprograms/scent_free.html [Accessed 22 Sep. 23].

(20) Singal, M. et al. (2011). Fragranced products and VOCs. *Environmental Health Perspective*, 119(5), A200.

(21) Caress, S.M., and Steinemann, A. C. (2009). Prevalence of fragrance sensitivity in the American population. *Journal of Environmental Health*, 71(7), 46–50.

(22) International Agency for Research on Cancer. d-Limonene. IARC Monogr Eval Carcinog Risk Hum, 73, 307–327. 1999. Available at: www.inchem.org/documents/iarc/vol73/73-11.html [Accessed 22 Sep. 23].

(23) Radis-Baptista, G. (2023). 'Do synthetic fragrances in personal care and household products impact indoor air quality and pose health risks.' *Journal of Xenobiotics,* 13(1): 121–131.

(24) King, A. (2017). Cosmetic and perfume sales staff exposed to high phthalate levels. Available at: www.chemistryworld.com/news/cosmetic-and-perfume-sales-staff-exposed-to-high-phthalate-levels/3008323.article [Accessed 23 Sep. 23].

(25) Royal Society of Chemistry UK's Paint Stash Totals over 50 Million Litres, Posing Environmental Issues. (n.d.). Available at: www.rsc.org/news-events/articles/2022/jan/uks-paint-stash-totals-over-50-million-litres-posing-environmental-issues/ [Accessed 23 Sep. 23].

(26) American Lung Association. (n.d). Cleaning Supplies and Household Chemicals. Available at: www.lung.org/clean-air/at-home/indoor-air-pollutants/cleaning-supplies-household- [Accessed 23 Sep. 23].

(27) California Air Resources Board. (2020). Volatile Chemicals in Cleaning Products Can Cause Indoor Air Pollution. Available at: ww2.arb.ca.gov/resources/fact-sheets/cleaning-products-indoor-air-quality [Accessed 23 Sep. 23].

(28) COSHH and Cleaners Key Messages. (n.d). Available at: www.hse.gov.uk/coshh/industry/cleaning.htm [Accessed 23 Sep. 23].

(29) Consumer Product Safety: Advice for Staying Safe. (2019). Available at: www.gov.uk/guidance/consumer-product-safety-advice-for-staying-safe [Accessed 23 Sep. 23].

(30) Ventilation and Filtration. (2022). Available at: www.evotechairquality.co.uk/articles/ventilation-filtration [Accessed 23 Sep. 23].

(31) Looi, M. K. (2023). How legionnaires' disease evacuated an asylum seekers' boat. *BMJ*, 382. https://doi.org/10.1136/bmj.p1876

(32) Baker, N. and Steemers, K. (2019). *Healthy Homes*. London: RIBA Publishing.

(33) CIBSE KS17. (2011). Indoor Air Quality & Ventilation. Available at: www.cibse.org/knowledge-research/knowledge-portal/ks17-indoor-air-quality-ventilation [Accessed 23 Sep. 23].

(34) ANSI/ASHRAE Standard 62.2-2022 Ventilation and Acceptable Indoor Air Quality in Residential Buildings. Available at: www.ashrae.org/technical-resources/standards-and-guidelines/read-only-versions-of-ashrae-standards [Accessed 29 March 2019].

(35) ASHRAE Position Document on Indoor Air Quality: Approved by ASHRAE Board of Directors July 1. Available at: www.ashrae.org/file%20library/about/position%20documents/pd_indoor-air-quality-2020-07-01.pdf [Accessed 23 Sep. 23].

(36) BEAMA Guidance for Improving Indoor Air Quality in Existing Homes (2023). Available at: www.beama.org.uk/resourceLibrary/guidance-for-improving-indoor-air-quality-in-existing-homes.html [Accessed 23 Sep. 23].

(37) AECOM Ltd. (2019). Ventilation and Indoor Air Quality in New Homes. Ministry and Housing Communities and Local Government September 2019.

(38) Marrs, C. (2016). Airtightness blamed for health risks in homes. *Architects Journal*, November 1, 2016. Available at: www.architectsjournal.co.uk/buildings/airtightness-blamed-for-health-risks-in-homes [Accessed 23 Sep. 23].

(39) Taylor, M. and Morgan, L. (2011). Ventilation and Good Indoor Air Quality in Low Energy Homes Good Homes Alliance November 2011. Available at: https://goodhomes.org.uk/wp-content/uploads/2017/08/VIAQ-final-120220.pdf [Accessed 23 Sep. 23].

(40) Public Health England Air Quality: UK Guidelines for Volatile Organic Compounds in Indoor Spaces. (2019). Available at: www.gov.uk/government/publications/air-quality-uk-guidelines-for-volatile-organic-compounds-in-indoor-spaces [Accessed 23 Sep. 23].
(41) Department of Finance NI. (2023). Review of energy efficiency requirements and related areas of Building Regulations Discussion Document and Pre-consultation on next steps. Department of Finance Northern Ireland 26 July 2023. Available at: https://www.finance-ni.gov.uk/news/views-sought-energy-efficiency-and-related-areas-building-regulations [Accessed 29 March 2019].
(42) Zero Carbon Hub, NHBC Foundation Mechanical Ventilation with Heat Recovery in new Homes Final Report. (2013). Available at: www.nhbcfoundation.org/wp-content/uploads/2016/05/RR8-Mechanical-ventilation-with-heat-recovery-in-new-homes.pdf [Accessed 23 Sep. 23].
(43) Sharpe T. (2022). Delivering Healthy Homes: Indoor Air Quality with Ventilation. Available at: https://goodhomes.org.uk/events/delivering-healthy-homes-online-tutorial-series-iaq-ventilation(https://goodhomes.org.uk/wp-content/uploads/2017/08/VIAQ-final-120220.pdf [Accessed 23 Sep. 23].
(44) Aereco Downloads. Available at: www.aereco.co.uk/downloads [Accessed 23 Sep. 23].
(45) Indoor Air Can be 5x more polluted than outside. Available at: www.vent-axia.com/healthyhomes [Accessed 23 Sep. 23].
(46) Homeowners Fed up with Condensation Mould or Damp? Available at: www.envirovent.com/sectors/private-homeowner/ [Accessed 23 Sep. 23].
(47) Abrams, D.S., op cit. An evaluation of the effectiveness of a recirculating hood. *American Industrial Hygiene Association Journal*, 47(1), 181986
(48) Jacobs, P., and Borsboom W. (2017). Available at: www.aivc.org/resource/efficiency-recirculation-hoods?collection=34627 [Accessed 23 Sep. 23].
(49) Kumar, P. et al (2023). Mitigating exposure to cooking emissions in kitchens of low-middle income homes. A guide for home occupants, owners, builders and local councils. Global Centre for Clean Air Research Guildford Living Lab University of Surrey 2023.
(50) Corinne Get Rid of New House VOC Offgassing Odor. (2019). Available at: www.mychemicalfreehouse.net/2019/05/mitigating-sealing-remediating.html [Accessed 23 Sep. 23].
(51) Pinto, M. A. (2021). Building a Bake-Out. www.randrmagonline.com/articles/89283-building-a-bake-out-revisiting-the-basics-of-an-effective-chemical-remediation-tool [Accessed 23 Sep. 23].
(52) Morantes, G., Jones, B., Molina, C. and Sherman, M. H. (2023). Harm from indoor air contaminants, *SSRN*. Available at: https://papers.ssrn.com/sol3/papers.cfm?abstract_id=4409736 [Accessed 23 Sep. 23].
(53) Jones, B. (2023). Dallying with DALYs. *ASHRAE Journal*.
(54) Morantes, G., Jones, B., Sherman, M. H. and Molina, C. (2023). A preliminary assessment of the health impacts of indoor air contaminants determined using the DALY metric. *International Journal of Ventilation*. https://doi.org/10.1080/14733315.2023.2198800 [Accessed 23 Sep. 23].
(55) Improving Indoor Air Quality. (n.d). Available at: www.epa.gov/indoor-air-quality-iaq/improving-indoor-air-quality [Accessed 23 Sep. 23].
(56) Kukadia, V. and Upton, S. (2019). Ensuring good indoor air quality in buildings BRE Trust. Available at: www.bregroup.com/bretrust/wp-content/uploads/sites/12/2019/03/Ensuring-Good-IAQ-in-Buildings-Trust-report_compressed-2.pdf [Accessed 23 Sep. 23].
(57) Medicair Stop Mould Spores Spreading. (n.d). Available at: ww.medicair.co.uk/pollutants/air-purifier-for-mould/ [Accessed 23 Sep. 23].
(58) Airmid Health Group. (n.d). Available at: https://airmidhealthgroup.com/ [Accessed 23 Sep. 23].
(59) Wargocki., P. (2013). 'The effects of ventilation in homes on health.' *International Journal of Ventilation*, 12(2), 101–118.
(60) Coules, C. (2022). Ventilation reduces indoor pollution levels except for NO2. Available at: https://airqualitynews.com/headlines/ventilation-reduces-indoor-pollution-levels-except-for-no2-study-finds/ [Accessed 23 Sep. 23].
(61) Moreno-Rangel, A. (2023). Indoor Air Quality and Thermal Environment Assessment of Scottish Homes with Different Building Fabrics MDPI. Available at: www.mdpi.com/2075-5309/13/6/1518) [Accessed 23 Sep. 23].

(62) McGill, G. (2015). An Investigation of indoor air quality in UK Passivhaus dwellings. In: Smart Energy Systems and Buildings for a Sustainable Future. Springer.
(63) McGill, G. (2015). Bedroom environmental conditions in airtight mechanically ventilated dwellings. In: Healthy Buildings Conference, Europe, 18-20th May, Eindhoven.
(64) McGill, G., Moore, J., Sharpe, T., Downey, D. and Oyedele, L. (2015). Airborne bacteria and fungi concentrations in airtight contemporary dwellings. *Healthy Buildings America*. Available at: www.mendeley.com/research/airborne-bacteria-fungi-concentrations-airtight-contemporary-dwellings [Accessed 27 Mar. 24].
(65) Sharpe, T., and Morgan, C. (2014). Towards low carbon homes – measured performance of four passivhaus projects in Scotland. In: Eurosun 2014, 16–19 September 2014, Aix-les-bains, France.
(66) Sharpe, T., Gupta, R. and Mawditt, I. (2016). Characteristics and performance of MVHR systems: A meta study of MVHR systems used in the Innovate UK Building Performance Evaluation Programme. Available at: www.researchgate.net/publication/296195515 [Accessed 23 Sep. 23].
(67) Howieson, S. G., Sharpe, T. and Farren, P. (2013). Building tight – ventilating right? How are new air tightness standards affecting indoor air quality in dwellings? *Building Services Engineering Research and Technology*, 35(5), 475–487. ISSN 0143-6244.
(68) Sharpe, T., Porteous, C. D. A., Foster, J. and Shearer, D. (2014). An assessment of environmental conditions in bedrooms of contemporary low energy houses in Scotland. *Indoor and Built Environment*, 23(3).
(69) Sharpe, T. (2014). Investigation of Occupier Influence on Indoor Air Quality in Dwellings. Technical Report Scottish Government.
(70) Morgan, C., John Gilbert Architects. (2021). Indoor Air Quality in Airtight Homes A Designers Guide. The HEMAC Report. Available at: www.seda.uk.net/products/q1urmll4rtkao5add1cyzk3wdjc6hu#:~:text=Airtight%20homes%20can%20overheat%2C%20become,with%20respiratory%20or%20immunity%20problems [Accessed 23 Sep. 23].
(71) Shurbsole, C., Macmillan, A., Davies, M. and May, N. (2014). 100 Unintended consequences of policies to improve the energy efficiency of the UK housing stock. *Indoor and Built Environment*, 23(3). Available at: https://doi.org/10.1177/1420326X14524586 [Accessed 27 Mar. 24].
(72) Jiang, S-Y., Ma, A. and Ramachandran, S.(2018). Negative air ions and their effects on human health and air quality improvement. *International Journal of Molecular Sciences,* 19(10): 2966.
(73) Luo, E. (2019). The Effect of Negative Ions. Available at: www.healthline.com/health/negative-ions [Accessed 23 Sep. 23].
(74) UKGBC Sustainable Pre-insulated Air Duct. (2022). Available at: https://ukgbc.org/resources/ecoduct/ [Accessed 23 Sep. 23].
(75) Ecoduct Product Summary and Technical Data sheet. Available at: www.ecoduct.co.uk/documentation [Accessed 23 Sep. 23].
(76) Kooltherm. (2023). Duct Insulation and the Koolduct system Product Safety Information. Available at: www.kingspan.com/gb/en/products/hvac-insulation/duct-insulation/kooltherm-duct-insulation/ [Accessed 23 Sep. 23].
(77) Hamid, O. (August 7, 2023). Personal email from the UKGBC communications and Media Manager.
(78) Technical Booklet K Building Regulations NI. Available at: www.finance-ni.gov.uk/publications/technical-booklet-k [Accessed 23 Sep. 23].
(79) Shrubsole, C., Dimitroulopoulou, S., Foxall, K., Gadeberg, B. and Doutsi, A. (2019). IAQ guidelines for selected volatile organic compounds (VOCs) in the UK. *Building and Environment*. 165, 106.

7 Testing for indoor air quality

While it is a requirement of the building regulations to test for air tightness, the effectiveness of the ventilation system is not checked, and while there are limits set out in the regulations for TVOCs, no tests are done to check this either. However, there are building owners, householders, architects, specifiers and builders who may want to check indoor pollution levels. On occasion the author has had an opportunity to do this with other architects and their clients. In one project, at a hospice in England, VOC and formaldehyde emissions were checked in an existing building, prior to renovation works being carried out. The building was to be totally renovated and extended and advice was given to the architects about what materials to avoid in the new construction works. A further emissions check was carried shortly after works were complete, when VOC emissions might have been expected to be higher, but the indoor air quality was found to have improved. This could have been the result of selecting low emissions products instead of conventional materials, but it was not possible to do further research or publish the results as this was blocked, due to opposition from a member of the hospice board.

Carrying out such indoor air quality testing can be a useful tool to check whether good or bad air quality is present in buildings, but it is rarely done. In many cases testing is only carried out when occupants become seriously worried about their health, being affected by mystery odours, or something they cannot put their finger on. There are a range of options for carrying out indoor air testing but many can be quite expensive. Some universities may be able to offer indoor air testing, but surprisingly, where extensive work on outdoor air emissions is available, using advanced technology, some university departments have little knowledge or experience testing indoor air quality. A web search for universities to carry out indoor air quality testing will yield few results apart from the University of Chester, but disappointingly they are yet another body who are testing for emissions from cooking and cleaning only.

> The project aims to understand the sources and reactions of pollutants indoors. Scientists will investigate the emissions from different pollution sources indoors, both in terms of their strength and their composition; and examine the key chemical reactions that occur indoors following these emissions, together with the harmful products formed.
>
> As part of the cooking aspect of the research, the team will be cooking an agreed set of standard household recipes, on a regular basis, to monitor the impact on air quality within the unit. **(1)**

There are a few private consultancy organisations that are able to do this work. Much of the bread and butter work for such consultancies is testing workplace emissions and places like hospital operating theatres, where it is critical that ventilation systems, filtering and emissions

DOI: 10.1201/9781003129226-7

are under control. This kind of work can be quite expensive and there is little interest in doing small-scale domestic testing.

Companies that carry out indoor air quality sampling may use "sorbent" tubes for active (pumped) sampling. The sampling may be carried out in compliance with ISO 16000-6 or US EPA Method TO-17. Sorbents are also used for passive sampling methods which generally require longer sampling times. Equipment for this kind of sampling can be bought from Markes International in the UK, but they do not carry out air sampling analysis themselves. **(2)** Markes were asked to suggest organisations that could carry out indoor air quality sampling, using their equipment, but their suggestions only included the Health and Safety Executive Northumbria Waste and a couple of commercial organisations that said they were not interested in the domestic market. The fees set out on these company websites are very high. Fortunately, it is possible to carry out air sampling tests at an affordable price, but the analysis is done by laboratories in the USA and Denmark, rather than the UK. Details are discussed below.

Low-cost sensors

Reference has been made in other chapters to the UK Government and funded research projects pre-occupation with "low-cost sensors." A large number of companies have entered the market offering such "low-costs sensors" and these are being used in a number of research projects discussed in Chapter 4. Much of this work is being carried out by organisations that come from a traffic pollution background, and have previously been pre-occupied with particulates. The low-cost sensors can give an indication of particulate levels, TVOCs, CO2, temperature, humidity and possibly formaldehyde. The author has a hand-held sensor that cost as little as £35 from a well-known online seller. It is little more than a toy, but it does give a hint of elevated TVOC, formaldehyde and CO2 levels. No information was provided with the sensor on how the device works. There are mobile phone apps that pretend to give information on air quality, but these need to operate with some kind of sensor.

So far, these products are not regulated, and how they are calibrated is something of a mystery. General TVOC readings are only useful in giving some indication that there might be a general problem with VOC emissions. Low-cost sensors do not give any detail as to the source of chemical emissions and provide no indication as what might be done in terms of remedial action. It is becoming commonplace to install these sensors in buildings to monitor air quality throughout the year, even though it is unusual for anyone to check or monitor results. As was seen in Chapter 4 highly funded "research" projects claim to be using low-cost sensors rather than using proper scientific analysis of emissions.

It is far better to carry out air sampling using much more detailed air sampling methods. This is done on a single day and can then be repeated weeks later. Leaving a sensor in a building over a period of time is not effective, except perhaps to give an indication of when chemicals have been brought into a building by the occupants, as this probably does not identify emissions from the building fabric.

In addition to air sampling, it is possible to collect dust samples. There is an innovative water sampling method developed in Finland that the author has not yet had an opportunity to examine. The samples must go to a laboratory that will analyse the results and provide details of the various pollutants detected.

How do low-cost VOC sensors work?

Some low-cost sensors use a heated film of metal oxide (MoX) particles. Adsorbed oxygen reacts with "target gases" which release electrons, thereby changing the electrical resistance of the metal oxide layer. The sensors are calibrated with a gas such as ethanol, which is the basis for estimating VOC concentrations such as n-octane and m-xylene. Low-cost sensors only estimate small groups of VOCs. **(3), (4)**

> In general, the main drawback of these devices is the lack of sensitivity and/or selectivity to benzene. Most of the MOx and amperometric OEM sensors are not able to reach levels lower than 100 ppb of benzene, although a few embedded sensors device show a sensitivity to few tens of ppb. Their limit of detection are two to three orders of magnitude too high for monitoring benzene in ambient air at the desired 1 ppb limit of detection. Portable gas chromatographs and ion-mobility spectrometers would be both sensitive and selective. Unfortunately, they are simply too expensive for the low-cost sensor market. **(5)**

Despite the inadequacy of low-cost sensors, they are being widely used by respected scientists to give the impression they are monitoring indoor air quality. Even DEFRA has warned of their limitations. **(6)**

Another technology, smart nanotubes (the "electronic nose"), is being introduced to detect gas leaks and VOCs.

> There are still many open questions, mainly related to the lack of openness and standardization of calibration and analysis procedures, evaluation of performance, handling and quantification of interferences. **(7)**

The ready acceptance of low-cost sensor technology indicates an unwillingness to carry out thorough scientific testing of building chemical emissions. It is driven by an assumption that indoor air quality is merely a problem of cooking, wood burning stoves, domestic cleaning and cosmetic goods. There may well be a role for low-cost sensors, but a rigorous approach would be to examine and test these first, before using them to collect data. Some low-cost sensor product companies were contacted with requests for more detail but were unwilling to share any information about whether they had been independently tested.

Using effective air sampling

Given the inadequacy and expense of many air sampling methods, it is fortunate that access to affordable air sampling is available from Waverton Analytics, based in Cheshire, England. Waverton provides a Home IAQ survey method which involves placing a small pump with stainless steel thermal desorption tubes that can detect a range of VOCs, formaldehyde and mould metabolites (MVOCs). One test can cover an area of 200 square metres with internal doors left open. **(8)**

The tests and laboratory analyses can be carried out for a few hundred pounds, unlike the thousands charged by some Universities and UK commercial laboratories. The sampling tubes are sent to a laboratory in the United States run by a company called Enthalpy Analytical, and they send back detailed information on the levels of VOCs detected. (Enthalpy is a wholly owned subsidiary of Montrose Environmental.) The method used is regulated by the USA Environmental Protection Agency (EPA).

> When it comes to the sampling and analysis of volatile organic compounds (VOCs) in ambient/indoor air, and those entrained in contaminated soils/underground water table, EPA Compendium Method TO-15 is the most used analytical method in the United States. TO-15 is the EPA's Selected Analytical Method (SAM) for air samples for environmental remediation and recovery.

"TO" in TO-15 stands for toxic organics. This method documents sampling and analytical procedures for the measurement of subsets of the 97 VOCs that are included in the 189 hazardous air pollutants (HAPs) listed in Title III of the Clean Air Act Amendments of 1990. Full list of the 97 VOCs may be found in Table 1 of the Method TO-15.

Analytical Instrumentation for TO-15 analysis

Laboratory analysis of the samples involves use of a pre-concentrator to focus small amounts of VOCs from large volumes of air. A gas chromatograph is then used to separate the individual VOC components and a mass spectrometer is used to identify and quantify each individual component in the sample. **(9)**

The easiest way to give an indication of what can be obtained for the Waverton IAQ home survey is to summarise the results from a typical report. A typical report summary is shown in Figure 7.1.

The report shows in red severe levels of VOCs in the property tested. In the case shown in Figure 7.1 a total VOC level (TVOC) of 3,600 µg/m³ is declared. The report, which normally runs to 12 pages, gives more detail as to the likely source of these emissions. This report also indicates moderate levels of light hydrocarbons and moderate levels of lifestyle-related products. Mould levels in this case were moderate at 14 µg/m³. What is useful about the report is that it compares emission levels found in the house to a contamination index based on results in other

Your Indoor Air Quality Report Summary

Your Indoor Air Quality Report has several sections describing different aspects of your home's air quality. A summary of this data is provided below, additional information and descriptions are included in the full report.

Total Volatile Organic Compounds (TVOC) Level

TVOC is a general indicator of the IAQ in your home (see page 2).

Total VOCs 3600 µg/m³

Total Mould Volatile Organic Compounds (TMVOC) Level

TMVOC is an assessment of the actively growing mould in your home (see page 3).

Total MVOCs 14 µg/m³

Contamination Index (CI) Level

The CI shows the types of air-contaminating products and materials that are present in your home (see pages 7, 8, and 9). These levels are estimates based on common home products and activities.

Building Related Sources

See page 7 for more detail.

H	Coatings (Paints, Varnishes, etc.)
N	PVC Cement
N	HFCs and CFCs (FreonsTM)

Mixed Building and Lifestyle Sources

See page 8 for more detail.

N	Building Materials-Toluene Based
N	Gasoline
N	Fuel Oil, Diesel Fuel, Kerosene
N	Moth Balls (Naphthalene Based)
N	Moth Crystals (p-Dichlorobenzene Based)
M	Light Hydrocarbons
N	Light Solvents
N	Methylene Chloride

Lifestyle Related Sources

See page 9 for more detail.

M	Personal Care Products
M	Alcohol Products
N	Odorants and Fragrances
N	Dry Cleaning Solvents
N	Medicinals

Note: Severity levels begin at Normal or Minimal and progress through Moderate, Elevated, High and/or Severe. The color progression from green to red indicates results that are increasingly atypical and suggest potentially higher risk.

Normal Moderate Elevated High Severe

Figure 7.1 A typical indoor air quality report

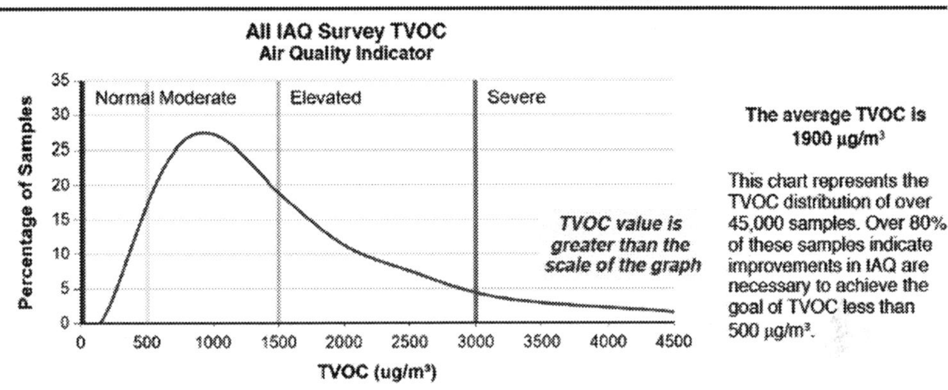

Figure 7.2 A typical volatile organic compound report

properties. In this case, the high VOC levels are attributed to paints, varnishes and sealants. This is probably due to decoration work having taken place recently. High levels of such decoration VOCs can mask emissions from other sources. The source of the light hydrocarbons may be due to building products or aerosols, natural gas, etc.

Emissions from personal care and cleaning products are found in most houses. In another report with even higher VOCs (4900 µg/m³) these were traced to coatings, gasoline, very high levels of light hydrocarbons and personal care and alcohol products (Figure 7.2). The reports are only able to list a range of possible sources, and it is essential for someone familiar with emissions to identify potential causes on a visit to and survey of the property. It might be easy to remove cleaning and paint products from the building, but many building products and decoration materials will already have been applied and cannot easily be removed.

TVOC figures from low-cost sensors may detect TVOC levels equivalent to this, but the air sampling report gives a great deal more detail. The air sampling method can provide information on a wider range of chemical emissions, but this increases the cost of the test.

The deniers of building materials chemical emissions will point to these reports and claim that this confirms that the emissions are caused by paints, decorations, cleaning and personal hygiene products as the main causes of bad indoor air quality, and they would be right. However, the health problems from these materials can be serious, and as they can be very high, this can mask lower level emissions from other materials. However, these tests can provide a clear indication of VOC and other contaminants, at a point in time, in a fair amount of detail. Removing paints, cleaning materials and personal hygiene products before a test is carried out makes it possible to identify emissions from the building more accurately.

It is also possible to carry out a surface dust sample which costs less than £200 at the moment. This involves the use of an adhesive slide and can be used to test for mould spores, pollen and other components of accumulated house dust. **(10)**

Mould data from air sampling in a house

When carrying out indoor air sampling on a brand-new housing association house in Northern Ireland emissions levels from recent decorating were surprisingly low. But it was discovered that the house had been left empty for 6 months or more. Despite this, VOC levels and formaldehyde readings were well over WHO safe levels. It is possible the house had not been ventilated whilst empty, so these chemical emissions were unable to dissipate. A surprising result was an elevated level of mould growth even though there was no visual evidence of this in what was a brand-new house. A thorough survey of the house was conducted after the test results were received and small traces of mould were found in the bathroom and an unoccupied bedroom. The tenants had only been in residence for a few weeks. The only ventilation was a totally inadequate small extract fan in the bathroom which came on when the light was switched on. The inevitable conclusion was that the moisture which led to the mould was effectively built into the construction of the house, and also due to the high relative ambient humidity, resulting from the conventional airtight construction. The tenants were unable to raise their concerns with the landlord, as they were on probation for a year, before getting a permanent tenancy (a normal practice apparently) and causing a fuss would have been a black mark against them! However, they were able to benefit from advice to improve ventilation and how to clean off the mould.

A problem in referring to air sampling reports of this nature is that many are done for private clients and so the full report remains confidential. A report was done on a private house in a small town for a client who was convinced that her asthma was caused by a being near to a main road. The house had moderately high VOC emissions, 940 ug/m3. The mould level was very low, as this was a relatively new, but well-ventilated house. In the more detailed breakdown Toluene and Styrene VOC levels were elevated but possible pollutants from traffic were very low. Identifying the source of VOC emissions requires a good understanding of building construction and materials. It is likely that the timber frame house, with stone and brick cladding, had been insulated with polystyrene with only a thin layer of plasterboard between it and the internal air.

In another study of an office, most VOC emissions were low and normal but a very high level of Toluene was detected. Toluene sources could be gasoline (petrol), adhesives, contact cement, solvent and heavy duty cleaner. Petrol emissions can get into internal air from linked garages or parking spaces outside the windows but other internal activities could also be to blame. The following is information from Enthalpy about VOCs that might feature in a typical report.

Significant VOCs

Based upon your specific home air analysis, the chemical compounds listed in Table 7.1 are significant contributors to the TVOC level reported on page 2 of your IAQ Home Survey Report or are indicative of specific types of products or problems. Compounds from a variety of chemical classes are represented here, although only the most common or most notable are specifically listed. These chemical compounds may come from a variety of sources as shown in the Contamination Index section of this report.

Locating and removing the source of the chemical compound is the most effective way to reduce the concentration of that chemical compound. If removing the source is not possible,

Table 7.1 Possible sources of emissions identified in a range of Waverton/Enthalpy Test reports

Ethanol	64-17-5	170	87	consumable alcohol; some solvents; renewable gasoline component; pharmaceuticals
Butane (C 4)	106-97-8	98	41	Aerosol propellant; cooking/camping/lighters fluids; liquefied petroleum gas (LPG); refrigerant; food additive
Pentane (C 5)	109-66-0	84	28	Aerosol propellant; blowing agent; gasoline fuel component
Isobutane	75-28-5	64	27	Gasoline and fuel additive; aerosol propellant; refrigerant; cooking/camping/lighter fluids
Acetone	67-64-1	55	23	Personal care, especially nail care; cleaners; paints and coatings; strippers and thinners; PVC cleaner; caulks and adhesives; wood filler; solvent
Cyclohexane	110-82-7	48	14	Solvent; glues and adhesives; some paints and coatings, petroleum fuel component
a-Pinene	80-56-8	48	8	Pine lumber; fragrances and essential oils; solvents; insecticides
Limonene	138-86-3 or 5989-27-5	39	7	Limonene (CAS 138-86-3) or d-Limonene (CAS 5989-27-5) Fragrances; paints and coatings; cleaners; solvent; preservative
3-Methylhexane	589-34-4	13	3	Adhesive; paints and coatings, petroleum fuel component
Hexane (C 6)	110-54-3	13	4	Solvent; adhesive; grease; lubricant; paints and coatings; petroleum fuel component
Toluene	108-88-3	12	3	Gasoline; adhesives (building and arts/crafts); contact cement; solvent; heavy duty cleaner

try to contain it in some way (e.g., placing the source in an air-tight container when not in use). In addition, many homes have insufficient ventilation so increasing the amount of outside air or filtering or purifying re-circulated inside air will almost always reduce the TVOC. Since VOCs may continue to off-gas even when the sources are stored, ventilation and air-purification methods may need to be employed continuously in order to maintain the VOC levels. **(10)**

Mould testing from Denmark

An alternative "low-cost: laboratory in Denmark can be used to check mould levels. House Test uses DNA analysis to diagnose mould problems using dust sampling or air testing. The tests are analysed using the qPCR method, known through Covid testing.

> Customers collect dust from a flat surface such as a door frame using a swab. The sample is then shipped to HouseTest's laboratory for analysis. Taking a sample is a simple process that requires no specialist training or equipment. This is the most frequently used of HouseTest's DNA analyses and is used for screening a building for concealed mould growth. The dust test can be used by specialists and private individuals alike. **(11)**

Denying emissions from building materials is easy when data is not available as evidence. Comprehensive studies have been carried out in Germany, France and Scandinavia where data has been collected on indoor air quality emissions. These can uncover serious hazards from

Table 7.2 Typical list of mould details from Housetest

The term universal fungi states the number of mould spores in the sample.
Universal fungi 43.022
Acremonium strictum 0
Alternaria alternata 2
Aspergillus fumigatus 2
Aspergillus glaucus grp. 2
Aspergillus niger 1
Aspergillus versicolor 1.555
Chaetomium globosum 3
Cladosporium cladosporides 343
Cladosporium herbarum 80
Cladosporium sphaerospermum 660
Mucor/Rhizopus grp. 49
Pen/Asp/Pae grp. 7.880
Penicillium chrysogenum 140
Penicillium expansum 2
Rhizopus stolonifer 1
Stachybotrys chartarum 17
Streptomyces spp. *182*
Tricoderma viride 6
Ulocladium chartarum 0
Wallemia sebi 93
These details are given for each dust swab location

dangerous substances but more sensitive analysis is required to identify the background levels of persistent endocrine-disrupting flame retardants and PFAS, for instance. It is frustrating to see the funding being squandered on low-cost sensors to detect cooking smells, rather than proper testing. Chemical analysis and wide-ranging collection of data on indoor air quality should be undertaken in the UK using scientific standards from the USA or Europe. This will collect data from cooking and wood burning stoves, but should not be biased towards this and provide much wider data on chemicals and mould that is caused by the building materials and construction.

References

(1) University project to analyses air quality in the home. Available at: www1.chester.ac.uk/news/university-project-analyse-air-quality-home [Accessed 21 Sep. 23].

(2) Indoor Air Quality Monitoring releases of pollutants into indoor air. Available at: https://markes.com/applications/environmental-monitoring/indoor-air/indoor-air-quality [Accessed 21 Sep. 23].

(3) How does Amotube VOC sensor work? Available at: https://atmotube.com/atmotube-support/how-does-atmotube-voc-sensor-work [Accessed 21 Sep. 23].

(4) Collier-Oxandale, A. M., Thorson, J., Halliday, H., Milford, J., and Hannigan, M. (2019). Understanding the ability of low-cost MOx sensors to quantify ambient VOCs. *Atmospheric Measurement Techniques*, 12, 1441–1460. Available at: https://doi.org/10.5194/amt-12-1441-2019 [Accessed 21 Sep. 23].

(5) Spinelle, L., Gerboles, M., Kok, G., Persijn, S. and Sauerwald, T. (2017). Review of portable and low-cost sensors for the ambient air monitoring of benzene and other volatile organic compounds. *Sensors (Basel)*, 17(7), 1520.

(6) UK Air. Low Cost Sensors – understanding the uncertainties. Available at: https://uk-air.defra.gov.uk/research/aqeg/pollution-sensors/understanding-uncertainties.php [Accessed 4 Oct. 23].

(7) Garcia, M. R., Spinazzé, A., Branco, P. T. B. S., Borghi, F., Villena, G., Cattaneo, A., et al. (2022). Review of low-cost sensors for indoor air quality. *Applied Spectroscopy Reviews*, 747–779. Available at: www.tandfonline.com/doi/full/10.1080/05704928.2022.2085734 [Accessed 21 Sep. 23].
(8) IAQ Home Survey. Available at: www.waverton-iaq.com/ [Accessed 21 Sep. 23].
(9) Ramzam, M. (2022). EPA Method TO-15 Sampling and Analysis 12.12 2022. Available at: https://enthalpy.com/blog/epa-method-to-15-sampling-and-analysis-fitting-your-needs/ [Accessed 21 Sep. 23].
(10) Surface Dust. Available at: www.homeaircheck.co.uk/surface-dust/ [Accessed 21 Sep. 23].
(11) IAQ Home survey op cit.

8 Healthy building, greenwashing and wellwashing

Just what were you expecting when you chose a petrochemical over a natural insulation, like cellulose, wondered Robert Riversong?

> As I've stated more times than anyone here cares to remember, there is nothing either 'green' or healthy about petrochemical foams, or any of the 80,000 petrochemicals that never existed on earth before we created them.
>
> Unfortunately, most of American society is brainwashed into believing in the 'magic' of chemistry, as the advertisers and marketers have impressed on us for generations. Every product produced since the start of the petrochemical age is toxic, either to people or the environment or both.
>
> In spite of this growing understanding, most professionals in the "green" building movement continue to rationalize the use of petrochemical materials in order to save petrochemical energy, further rationalized by the belief that it makes a house durable. **(1)**

In this chapter, the options available to reduce chemical emissions and create healthy homes and buildings are discussed. This chapter has turned out more critical than first intended as it is necessary to deal with the problem of greenwashing and wellwashing. Even though scientists, academics and Government are struggling to cope with the problem of hazardous building materials, many ordinary people have worked out the dangers for themselves and are looking for alternative and healthier solutions. This is not easy if they go to their local DIY superstore or builder's merchants as they will not get very far trying to source non-toxic and healthier products. They will also have problems if they go to an architect who tries to persuade them to go for a passivhaus low-energy building, constructed from petrochemical and other hazardous insulation materials. Even those architects who are interested in greener solutions may not have enough knowledge about alternative materials and how to use them.

The interest, if not the knowledge, is there, as evidenced by the 240 architects and others who attended a webinar organised by the Green Register on natural materials given by this author in early 2023 but healthier buildings remains a fringe concern for the mainstream construction industry. **(2)**

There are a range of campaigning "green" organisations that unfortunately cannot provide good advice and information as they are confused about the issues and solutions. The example already cited in Chapter 6 of the IGBC, apparently promoting a duct system made from Phenolic, and then denying that they are endorsing this, and many other products is typical. Organisations like the IGBC and the World Green Building Council, while claiming to be green, are working with and relying on funding from the mainstream construction and

DOI: 10.1201/9781003129226-8

materials industry. Multi-national companies that have managed to block progressive change in the European Union, as they want to continue to sell their climate and health-damaging products, support a range of greenwash organisations. Stick the label Eco, Sustainable or Green at the front of the name of a product and it will not be challenged by the Advertising Standards Authority as they are also confused about how to interpret environmental issues.

Several initiatives such as Environmental Product Declarations, Health product declarations, circular economy certification, low or net or zero carbon claims and many more look good in principle but are manipulated by mainstream industry to ensure that they can carry on selling the same old stuff! They even develop sophisticated computer programmes and incorporate these measures into Building Information Modelling (BIM) without it necessarily changing the nature of the building. Perhaps the most significant recent change is moving towards counting embodied energy. Embodied energy is the assessment of the amount of fossil fuel energy used to create, transport and install a product. This was discussed in the Green Building Handbook in the early 90s, but mainstream industry has only just caught up. **(3)**

Government, local authorities and private developers are now requiring embodied energy to be taken into account in their procurement and construction policies. Happy to oblige, mainstream petrochemical materials companies are now producing glossy embodied energy guides, while barely changing their products or reducing the embodied energy in their products. However, there is a growth of production of bio-based, natural and low embodied energy materials but these still remain marginal in their use.

Bio-based natural materials also provide the backbone of healthier chemical-free alternatives and some big multi-nationals are encroaching on this territory, but when this happens they increase the cement or chemical and glue content of previously natural materials to cut costs and increase profits. Meanwhile mainstream builders and their clients are very nervous about departing from standard construction approaches, and while they espouse green policies, they fail to implement them in their commissioning. The manufacture and supply of natural, chemical-free or low-chemical products are increasing rapidly but they are still hard to source and architects are still nervous about specifying them. A guide to the range of natural and bio-based materials is available. **(4)**

Natural material deniers

Just as we have building emission deniers, there are those that attack natural chemical-free materials, or at least have a host of excuses why not to use them. The first line is that they are too expensive, which is true for a few materials, but not all. Some distributors do profiteer when selling natural materials, assuming that only wealthy "eco-style" people will be interested. You will be told that they are difficult to obtain, that supply chains are poor, transport costs are high, etc. A Polish self-builder in the West of Ireland, known to the author, found it cheaper to hire a truck and drive himself to Poland to source affordable natural materials than to order them through distributors in Ireland. However, as natural materials become more popular, obtaining them will be easier, but whether they are healthier is discussed in more detail below by examining different materials and products.

However, mainstream construction industry professionals continue to attack natural materials, with examples earlier in the book suggesting that breathability is a myth and that natural materials emit much higher VOCs.

The National House Building Council (NHBC) produced a report about natural materials in 2014 which attracted a great deal of negative publicity about natural materials.

For such materials to become more widely accepted, it is important that they can be properly evaluated and assessed alongside other products and materials.

This report provided a brief history of the subject, and reviewed the current developments in the use of cellulose-based building materials. It discussed use and performance issues and examined the potential benefits and risks associated with natural materials and provided examples of recent projects built in the UK, though it included thatching and straw while ignoring other more important bio-based materials.

> It is important to emphasise that the current low incidence of use of the materials featured in this report in built properties means there is, as yet, relatively little statistically robust evidential data as to their long-term performance. A primary purpose of the report, therefore, is its aim to assist those interested in pursuing the use of the featured materials in new homes and guiding them towards sources where further information and data may be obtained. **(5)**

Despite the fact that the report was referring to materials such as earth and thatch, which have been in use for hundreds of years, this doubt-raising comment was picked up by the Construction Press leading to headlines such as "Builders warned of risks with natural materials." **(6)**

So widely reported and influential was this misleading conclusion that it led indirectly to the collapse of several companies producing natural materials and much harder market conditions for the producers of natural healthy materials. Examples include Black Mountain Sheep Wool insulation in North Wales which closed down and then was taken over by International Petroleum Products. Excel Industries producing cellulose insulation in South Wales went into receivership and was taken over by CIUR in Prague. Natural Building Technologies (NBT) had to be rescued through a partnership with Pavatex, the Swiss Wood fibre insulation materials supplier. NBT was later taken over by Soprema, a French plastic roofing company.

Hempcrete or hemp lime, as discussed in more detail below, also suffered a major setback when mainstream organisations tried to build housing projects helped by significant Government funding. One of the projects led by Kevin McCloud, of TV programme Grand Designs fame, had a great deal of bad publicity in the construction press which set back the development of hempcrete for a number of years, due to problems with damp and drying out. **(7)**

Timber and solid wood

Timber is the most complex of natural materials to discuss as wood includes a range of natural oils, tannins, lignin and tiny amounts of formaldehyde, though this can vary considerably between different tree species. The general consensus is that VOC emissions are relatively low but some natural material deniers are quick to suggest that natural wood is toxic.

> Wood is a common natural product with a typical pleasant smell composed of main structural compounds of polysaccharides (cellulose, hemicelluloses, and lignin) that contain a wide range of low molecular weight organic chemicals and extractives… . …the wooden extractives can be divided into groups—lipophilic or hydrophilic (or polar) components. An important portion of wood extractives are volatile organic compounds (VOCs) formed by terpenes, terpenoids, flavonoids, alcohols, aldehydes, and ketones, also in smaller amounts of higher alkenes and fatty acids. This is a low, but still well detectable, amount of VOCs, that can be released from wood.

Various studies have analysed the range of VOCs emitted from natural timber including Limonene but the areas of concern are higher with the adhesives:

> In case of wood-based panels, the impact of adhesives and additives that are essentially applied aiming to adjust the panels' properties is even enriching this cocktail of chemicals. **(8)**

People have been living in wooden buildings for centuries without documented ill effects, but unfortunately the chemical industry has convinced many people that timber will rot and burn unless treated with preservatives, insecticides, flame retardants and so on. Chemically treated timber is popular with some architects, but since the Grenfell fire some have stopped specifying timber altogether as they think it will burn, even though timber was not a factor in the Grenfell fire. Timber cladding both internally and externally is viewed by some as a fire hazard and many building control staff insist that it is treated with intumescent paints. Intumescent paints usually contain a range of dangerous solvents and VOCS, and even when water-based, and are a risk to indoor air quality and health. **(9)**

Large sections of solid timber are not a fire risk as they should char rather than burn, bandut smaller sections of timber can burn but are less of a fire risk than the many plastic materials now used in buildings.

Timber frame construction

OSB

A great deal of modern construction by housing developers uses timber frames of various kinds, usually prefabricated. These timber panel buildings use sheathing boards of OSB (Oriented Strand Board) which is made with significant quantities of glues. The timber frame is usually wrapped in plastic air-tight membranes, filled with various synthetic, potentially flammable, insulation materials and lined with gypsum board internally. External cladding can be of brick and other materials. This is not the healthiest use of timber and some of these forms of construction are a fire risk. A typical health and safety data sheet for one of the best known OSBs contains the somewhat bizarre statement "Non-flammable at room temperature, **but will burn.**" **(10)**

Manufacturers will claim that OSB in panel form is unlikely to give rise to any toxicological effects; however, health risks may arise from dust and moulds associated with poor processing, handling or storage practices. OSB contains significant amounts of adhesives, but the chemical composition of these is rarely disclosed other than possible references to formaldehyde. The boards also receive a "coating" which is rarely specified either.

> Formaldehyde has a WEL of 2.5 mg/m3 (8 hour TWA) and a Short-Term Exposure Limit (STEL) also 2.5 mg/m3 (15-minute exposure). Formaldehyde vapour can irritate the eyes and nasal linings. Formaldehyde class for OSB is class E1 – less than or equal to 8 mg/100g (0.008 %) of board as per BS EN 13986:2004 Annex B. **(11)**

A leading brand of OSB is advertised as having zero **added** formaldehyde:

> Unlike traditional oriented strand board (OSB) containing urea-formaldehyde (UF) or melamine-urea-formaldehyde (MUF) binders, the Sterling OSB Zero family from Norbord uses a methylene diphenyl diisocyanate (poly-urethane MDI) resin to bind the thousands of strands that make up each board. **(12)**

This free-form formaldehyde statement is aimed to make it sound healthier and greener but according to the US EPA Diisocyanates in the glues are also dangerous.

> Diisocyanates are well known dermal and inhalation sensitizers in the workplace and have been documented to cause asthma, lung damage, and in severe cases, fatal reactions. **(13)**

Modern methods of construction

The UK Government has been promoting what is known as MMC (Modern Methods of Construction). The idea is that houses can be more efficiently prefabricated off-site. Most of the systems that have been developed are timber-based. Despite continued demands for more MMC, it has been a huge flop with millions of pounds wasted on collapsed MMC projects. Legal and General (L&G), one of the UK's biggest insurance companies, built a 55,000 square-metre factory near Selby, Yorkshire, to construct timber frame modular housing has made the majority of its 475 staff redundant, with losses of £174 million. The failed L&G project was initially set up to develop solid timber construction but abandoned this in favour of a cheaper approach. **(14)**

Many other MMC companies have also gone out of business, with big names such as Urban Splash, Countryside and many more. The Government and the construction industry is now bringing back concrete and have very short memories as they have forgotten the disastrous programme of system building in the 60s and 70s which led to the demolition of thousands of tower and deck access blocks built out of concrete. While MMC largely focuses on timber construction, concrete system building is making a comeback though several large concrete modular construction companies have also recently gone out of business. **(15)**

Surprisingly the technology of prefabricated concrete construction continues to have problems. Apart from possible low-level emissions from the cement additives and aggregates, buildings are failing, sometimes even before they have been occupied. Three recently constructed concrete modular buildings have been closed in Northampton and Somerset. The same company has worked on at least 15 other schools in the UK. There were concerns about the structural integrity of the buildings. The company concerned has gone into administration but bought up by another concrete company. **(16)**

It is important to discuss modular construction as it is frequently promoted as more eco-friendly and healthier even though there is little substance to such claims:

> Modular houses are often held to higher sustainable standards than regular housing, there's a focus on clean and green energy and using non-toxic or low VOC materials. **(17)**

Despite the many collapses of MMC companies, the UK Government continues to promote MMC as the best route to sustainable construction, and while MMC was intended to create bigger profits for house builders (through efficiencies) it was not really intended to create healthier homes. Despite this, some academics in the housing research world continue to promote MMC as the solution to housing supply issues and promote a misleading view of MMC as a route to better building quality.

> ...contemporary modern methods of construction that typify industry practice in the early 21st century do bring about **numerous benefits** for the development industry, one of which is **better building quality**. **(18)**

Payne and Serin's "evidence review" fails to address the risks and drawbacks of MMC. Written by lecturers in real estate and global urbanism they completely failed to address the issue of fire safety. UK National Fire Chiefs published a devastating critique of MMC in a 2022 report pointing out that MMC was not included in the new building safety act following Grenfell and pose many fire risks which have not been properly addressed. Most MMC timber frame systems are very lightweight, with only a thin layer of gypsum board separating chemical-emitting materials from the occupants. This increases the fire risk but also raises questions about the indoor air quality of such buildings. The concerns of the Fire Chiefs and other points were put to Payne and Serin in several emails over a 6-month period but they did not reply. The increased use of lightweight prefabricated buildings made of chemical materials is likely to emerge as a serious problem in years to come. **(19)**, **(20)**

Many MMC timber frame systems use flammable foam insulation materials, but even where safer materials are used, the panelised forms of construction make it much easier for fire to spread quickly. Two major fires, one in Shetland and the other on Fair Isle, in Scotland, where off-site MMC building system were used to build a hotel and a bird observatory, may provide clues as to the risks. Both projects were highly publicised during construction as prefabricated elements of the buildings had to be shipped from the mainland. **(21), (22)**

The Moorfield Hotel in Shetland and the bird observatory were both completely destroyed in a relatively short space of time. As is so often the case, the settlement of insurance claims means that the cause of fires and the inherent faults in building methods are rarely disclosed to the public. The trust that owns the bird observatory were contacted but refused to discuss the materials being used to construct the replacement building. The new replacement building also consists of timber off-site modules, constructed in Sheffield by IDMH Ltd. **(23), (24)**

There are a number of new companies developing off-site construction methods using timber frames and panels with natural, bio-based insulation materials such as hemp and lime or resins and hemp and wood fibre. It has not been possible to evaluate these systems as very little has been built. Most are forced to compromise using plastic membranes and glued timber boards, but these may be replaced by greener options. Thus, there is the possibility of healthier chemical-free MMC houses in the future.

Solid timber

Composite timber products such as plywood, OSB, I-beams, glulam, Laminated Veneer Lumber (LVL) and solid mass timber products involve the use of glues/adhesives. Solid timber (sometimes referred to as Cross-Laminated Timber [CLT]), in particular, is growing in popularity as it is seen as a more sustainable and healthier way to build. Unfortunately, the solid timber companies can be very evasive about adhesives and possible emissions. A guide to mass timber produced in Canada only mentions adhesives 5 times in 70 pages without going into any detail about the chemicals used. **(25)**

It was difficult to find research or publications on adhesives on the website of Napier University, the leading UK academic centre for research on composite wood. The mass timber industry seems more preoccupied with the effectiveness of adhesive types on compressive strength rather than indoor air emissions. **(26)**

Literature on chemicals in solid wood is sparse though there is research in progress to find alternatives such as soy-based adhesives.

> Each of the predominant commercial adhesives pose their own potential health risks. Engineered wood products containing MUF, MF, and other adhesives have been shown to

> emit formaldehyde during both production and use. Polyurethane-based materials, used in many building and furniture applications, potentially expose workers to toxic isocyanate compounds during production. In addition, polyurethane-based materials have an elevated fire toxicity compared to other commonly used building materials, primarily due to a greater emission of hydrogen cyanide (HCN). Therefore, there are potential health concerns associated with the use of several commercially available adhesives. **(27)**

Significant research and development work has also been done on adhesive-free composite timber developing Brettstapel and other approaches where the timber is pegged or screwed, but there has been limited takeup of this method. **(28)**

Wood fibre

A range of wood fibre insulation boards are available. These are generally made by pressing wood waste into panels. It is possible for the natural lignin in the wood to act as a binder and many wood fibre products do not contain chemicals or adhesives. Unfortunately, the industry is being taken over by bigger companies, some of whom are abandoning natural processes and using adhesives and flame retardants. Care should be taken when selecting this product to avoid the products with plastic and chemical additives. As is often the case it can be very difficult to track down the health and safety data sheet for some of these products. Wood fibre is generally regarded as a healthy material to use as it can provide air tightness, insulation and help to regulate humidity, but there is some debate about its flammability, and it can easily decay if it becomes too wet. While chemicals can be detected in climate chamber tests the chemicals emitted are at quite low levels.

> Using natural fibre insulation products, such as wood fibre, wool or hemp fibre can avoid health problems occurring in the installers and significantly reduce the amount of VOC's in the internal air of buildings, contributing to better indoor air quality. **(29)**

> The main components emitting from lightweight insulation fiberboards were acetic acid and aldehydes such as pentanal, hexanal, heptanal, octanal, nonanal, decanal, furfural, and benzaldehyde. **(30)**

VOC-free materials

It is possible to build most buildings using largely chemical-free materials. Even where chemicals are involved, these can be minimised with low VOC solutions. The reason that this does not happen in many buildings is due to ignorance and a cultural attitude within the construction industry that assumes that plastic and chemical-based materials are better. It is also due to laziness in terms of specification where architects and their colleagues cannot be bothered to look for alternatives. Organisations that promote low-carbon buildings like UKGBC, LETI and many more are happy to include chemical-based products in their standards, though if pressed, will say that they favour natural and ecological materials.

Building regulations and public procurement rules tend to favour products from the big commercial companies and demand certificates which smaller alternative companies cannot afford. A company like Clayworks, producing plasters and other materials from natural clay in the UK, are unusual for a natural materials business as they have grown massively and operate in the United States and the Middle East. They have been very successful at convincing high-end markets such as smart restaurants and West End shops to use natural healthy materials like clay and so they can afford to spend money on a range of certificates and approvals. However, much of the natural building materials industry in the UK remains a cottage industry, though there are signs that this is changing. **(31)**

Earth building, clay, earth and reed boards

Earth materials can be used for many purposes in buildings both structural and as finishes and are chemical-free and hygroscopic. Earth plasters are very popular in countries like Germany as a main contributor to healthier buildings, but are not so common in the UK.

> We can do a lot to avoid chemicals and plastics by seeking out natural materials and demanding transparency. Clayworks products have VOC emissions certificates and Health Product Declarations. Many other natural materials can be used within buildings for floors, wall build-ups, and insulation that can all contribute to keeping a space toxin free. Another way clay contributes to interior health is by absorbing and de-sorbing moisture–breathing. Unlike other finishes, clay does not 'cure': it remains a raw, porous material that constantly breathes, helping to maintain indoor air humidity at a level that promotes optimum health. **(32), (33)**

In addition to clay plasters and related materials, there is a range of boards made of clay and reeds and other materials. Products of this nature are widely used in Germany and other European countries but are uncommon in the UK. **(34)**

Lime and calcium silicate

Lime should always be preferred to cement for external renders and internal plasters. Lime and sand mixes are vapour permeable and can help masonry walls "breathe." Lime is also a natural biocide and can contribute to a healthy interior. It was used for thousands of years to reduce the risk of mould and infection from cow sheds to homes. You may not find that your local environmental health officer will understand this as they prefer plastic finishes that can be washed down with hazardous chemicals and disinfectants. There are a growing number of excellent suppliers of lime and lime-based products around the UK and demand for these materials is growing. There is a range of hydraulic, hydrated, hot lime and lime putty plus some chalk-based materials. Try to avoid the cheap hydrated limes from cement companies available in mainstream builders' merchants. The EU has recognised hydrated lime as a disinfectant and algaecide and the natural biocidal properties of lime are well understood if not widely applied. **(35), (36), (37)**

Straw

Strawbale building is popular with some ecobuilders, and there are many successful examples in the UK and around the world. Straw needs to be baled in small rectangular bales which are not always easy to get hold of, as farmers use large round bales today. Straw is sometimes mixed with earth to create cob walls but can also be laid in bales with a lime render. Strawbale building is much more difficult than is claimed by strawbale enthusiasts, and it is important to attend a good training course if you plan to use this. Strawbales can be susceptible to water and should be protected from driving rain by good roof overhangs. Straw is also susceptible to mould, and can cause a disease known as farmer's lung so great care should be taken to ensure that walls remain dry. **(38)**

Hempcrete, hemp blocks and hemp fibre

Hemp is the most versatile of plant-based (bio-based) materials for building. It is easy to grow but requires processing with specialist machinery. There are two processors in Yorkshire, and hemp is being grown throughout the UK and more companies are entering the market. Hemp is a very robust plant and is much less prone to damp and rot than straw. Hemp fibre can be used for a wide range of purposes from hemp resin composites as plastic replacement linings for aeroplanes and cars, to rope and fabric for clothes.

Hemp fibre can be used, sometimes mixed with flax and other crop-based fibres, to make insulation batts and quilts. These are now manufactured in Europe and in the UK and provide natural, healthy, robust insulation materials. Unfortunately, it is not always easy to find which products use any flame retardants or glues. The one product that does make a data sheet available reveals that apart from hemp it includes 20% recycled polyester and 10% polyester binder with an ammonium polyphosphate flame-retardant chemical, not an entirely natural material. Another hemp fibre insulation product only uses 10% polyester. Howeer, some of the hemp insulation materials are chemical-free, even if they don't make this clear in their marketing!

Hempcrete or hemp lime is made by mixed hemp straw (shiv or hurd) which is the woody core of the plant chopped into small pieces and then mixed with a lime binder and water and cast or sprayed into solid walls or blocks. It can also be used as a plaster or render and cast against existing walls. Currently hempcrete blocks are imported from various European countries, adding to transport costs, but the hemp grown and processed in the UK and lime binders made by UK suppliers can be used to make blocks or to cast into walls. Hempcrete can be mixed on-site and cast in shuttering or it can be sprayed using specialist machinery. Hempcrete does not include any chemicals or additives such as polyester but care should be taken to avoid some of the imported lime binders which contain high levels of cement. Binders with minimal cement are available.

Hempcrete provides very good thermal performance, with thermal mass that maintains a steady temperature inside buildings. It is also hygroscopic so it manages relative humidity and maintains a steady 50–55% relative humidity throughout the year. There are very few materials that can do this, though earth plasters are claimed to have a similar effect. Hempcrete walls are also vapour permeable (breathable) and as the lime used in the mix is a natural biocide, it's hard to think of a healthier material to maintain good indoor air quality. While lime is a chemical of a kind there should not be any hazardous emissions from hempcrete.

There is opposition to hemp products from some quarters due to its association with cannabis, but cannabis oil is growing in popularity and campaigns for the legalisation of medicinal cannabis production in the UK has parliamentary support. The development of hemp is still hindered by its licensing, controlled by the drugs section of the Home Office rather than it being accepted as an agricultural product.

The use of hygroscopic and vapour-permeable materials as a solution to mould problems is important to note. Hempcrete, in particular, has been found to maintain a steady level of humidity, known as buffering. It has been used in damp buildings to mitigate moisture and humidity and to work with existing brick and stone walls to maintain vapour open construction. A wide range of hemp and hempcrete building techniques are emerging including prefabricated panels. **(39), (40), (41)**

Sheep's wool

Huge quantities of fleeces from sheep go to waste every year. Farmers can barely cover the cost of shearing and so the wool gets buried, or even worse, burned. However, sheep's wool can make an excellent insulation quilt. Unfortunately, some products made in the UK also contain polyester, pesticides, borax and flame retardants as there is a worry about flammability and moth infestations. In fact, it's very hard to set fire to wool which is naturally flame retardant and there has been very little genuine evidence of moth infestations. It is possible to produce sheep's wool insulation which is chemical and biocide-free. A product made in Austria uses something called Plasma Ionic Protect and has almost negligible chemical input unlike the UK-made products. It is available in the UK. **(42)**

Cork and cork lime

There is a range of cork composite products. Cork trees do not grow in the UK, so it has to be imported from various European countries. Cork thermal insulating composites products can contain Hydraulic lime, Calcium dihydroxide, Silicic acid, sodium salt and potassium salt, but when adhesives are used they can contain a range of chemical additives. **(43)**

Coir, jute, cotton

There is a wide range of fibres available from tropical countries such as palm fibre, coconut fibre, jute and many more. An insulation product is being made in the UK from recycled jute from recycled coffee bags and waste woollen tweed material. An insulation made from cotton waste is also available. Some of these materials may contain polyester glues, moth proofing and flame retardants, and this information is not always easy to find, but a lot more use could be made of natural fibres from around the world which often end up as waste. **(44)**

Paints and finishes

It's not hard to find a level of complacency in indoor air quality literature about paints and finishes, as the increasing use of water rather than solvents in paints has undoubtedly reduced emissions. However, even water-based paints can be responsible for emissions that have negative health effects and cause allergies and bad reactions. Biocides are used as preservatives in paints such as 5-chloro-2-methyl-(2H)-isothiazol-3-0ne, 2-methyl-2(H)-isothiazol-3,one,1,2-benzisothiazol-3-(2H)-one and Bronopol. PFAS have also been introduced into so-called self-cleaning paints. Fungicides are also used in paints. While there has been a significant reduction in VOCs in paints as indicated by the labelling on paint tins, many paints still emit VOCs. **(45)**

It's also important not to assume that paints described as eco, natural or traditional are actually free of hazardous chemicals. Just because a well-known brand uses traditional colours that does not mean that it is greener or healthier, though many people seem to assume this is the case. Some eco paints are bogus whereas some really good paints made with earth and other natural materials may still contain very low levels of chemicals. Recent innovations include the use of graphene and nanoparticles in paints and this has raised some health concerns. Low odour wood stains have become popular and oils that penetrate wood can last much longer reducing the need to repaint every few years.

It's important to realise that most paints, even though they are water-based, are still made from petrochemicals and are plastic in nature as they are vinyl or acrylic, but there are alternatives to plastic paints using natural materials. **(46)**

Recycled cellulose insulation products that contain chemicals

It is important to examine data sheets to see what chemicals they contain. There is a range of recycled paper insulations, often referred to as natural, popular with ecoarchitects, but they can contain chemical binders, preservatives and fire retardants. About 20% of the contents can be Boric Acid, Magnesium Hydroxide, Calcium Sulphate and Calcium Carbonate. Some manufacturers are open about the chemical contents but others simply state "treated with inorganic flame retardants." **(47), (48)**

Recycled glass

A useful, largely chemical-free insulation product suitable for use in floors under floor slabs is recycled glass. This is made with a foaming process in products such as Glapor or Foamglass. These companies are decidedly cagey about giving details of their production process, but recycled glass is heated with a blowing agent near the melting point of the glass; the blowing agent releases a gas, producing a foaming effect in the glass. This is quite a different process than that used to make glass fibre and should have limited if any chemical emissions.

Mushroom insulation

An innovative material has emerged in recent years in which mycelium fungus is missed with various waste materials and allowed to grow. It is turned into a range of materials such as insulation batts. This is still at a very innovative stage and could lead to some exciting products, but given the concerns with mould discussed in this book it could take a while to gain acceptance!

Greenwashing

It is possible to construct buildings using natural, healthy and largely chemical-free materials, but to access information on how to do this it may necessary to wade through a jungle of misleading information about healthy building which can be summarised under the heading of Greenwashing or Wellwashing.

Greenwashing is recognised as a problem by many interested in sustainable construction, and there are several organisations set up to combat or expose greenwashing or to endorse green claims such as GUSTO which is supported by the Alliance for Sustainable Building Products. **(49)**

However, Gusto is a marketing company offering to help companies promote their products and only a handful of companies have so far signed their anti-greenwash charter at the time of writing. **(50)**

Biophilia

Biophilic design is promoted heavily by some architects and interior designers as a means to integrate building with nature and create healthy homes. Many people aspiring to a healthy building may find biophilic ideas inspirational, when a more natural approach is adopted for building design, but you will be lucky to find any serious attempt to reduce chemical emissions or improve indoor air quality in biophilic design guides such as in Kellert et al, with no mention of IAQ or emissions from materials in its 385 pages. There is talk of restorative environmental design and the use of natural materials, especially bringing plants into buildings, but not a mention of VOCs or hazardous chemicals. **(51)**

Oliver Heath, a well-known UK exponent of biophilia, does discuss VOCs, however.

> Choosing low or VOC-free materials for improved interior air quality (preferably with a Green premium ecolabel). VOCs (Volatile Organic Compounds) can have long term negative health effects if we breathe them in.

Sadly, there is no such thing as an independently certified Green Premium ecolabel other than a marketing device by Schneider Electric closely associated with Heath. **(52)**

Heath teams up with other commercial product companies such as Blue Air

Air Purifiers but his approach to "human centred design" does not seem to include much information about VOC emissions. **(53)**

Certification

There is a variety of green certification systems some of which make claims about healthy buildings but come close to another form of greenwashing called *Wellwashing*. A review of a wide range of systems warns that these can be seen more as eco-opportunism:

> The review finds that certifications can increase the perceived value of eco-friendly brands and consumer willingness to pay. However, the review also highlights the risks of greenwashing and free riding, which can undermine the intended benefits of certifications. Additionally, the institutional organization of certification systems may exhibit structural inertia, which may impede the integration of disruptive green technologies and market transitions ... the analyses conducted in the study indicate that while certifications can help prevent greenwashing, they can also contribute to eco-opportunism. **(54)**

One example of certification is the WELL building standard which provides an extensive level of requirements, but sets a poor level of acceptable VOCs (must be less than 500 µg/m3). It was developed by the American real estate industry to make commercial buildings appear healthier. The early version was very poor in terms of indoor air quality but it has improved a little over time. **(55)**

Even the relatively conservative RIBA is critical of WELL pointing out that BREEAM is much better than WELL and requires a lower VOC level of less than 300 µg/m3. **(56)**

Living building challenge

The Living Building Challenge (LBC) model created a great deal of interest when it was first launched but seems to have drifted into the background, at least in the UK. One company claims that LBC is "the most advanced measure of sustainability in the built environment." **(57), (58)**

The LBC system requires the testing of indoor air quality but seems to follow the poor WELL building standard of 500µg/m3 and formaldehyde level of 50 parts per billion which is well above the WHO safe level of 30. While it is encouraging that indoor air quality at least gets a mention LBC does not provide much detail about this. The LBC has a showpiece building in Seattle called the Bullitt building, which they claim to be the greenest building in the world, even though it is insulated with glass fibre and other rigid synthetic insulations materials, plastic membranes and asphalt roofing materials, not necessarily the greenest of choices. **(59), (60), (61)**

On the other hand, a recent publication by the International Living Future Institute entitled "The Regenerative Materials Movement" is an extensive source of information. However, it states that the LBC still allows "limited exceptions: for the use of foam insulation" while stating that "project teams should strive to eliminate these products from their specifications." The book contains a great deal of encouragement to use natural insulation materials. **(62)**

Building biology certification

Another form of healthy building certification which has had little impact in the UK is "Building Biology." Originating in Germany and backed by the Institute of Building Biology, it is probably the most comprehensive and holistic approach available. A small number of UK architects espouse this approach but some are also passivhaus exponents. Examples do not seem to use chemical-free materials.

Building Biology has 25 guiding principles which includes Healthy Indoor Air and listed as including:

- Supply sufficient fresh air and reduce air pollutants and irritants
- Avoid exposure to toxic moulds, yeasts and bacteria as well as dust and allergens
- Use materials with a pleasant or neutral smell
- Minimise exposure to electromagnetic fields and wireless radiation
- Use natural, non-toxic materials with the least amount of radioactivity

While they state that the goal of building biology is to create healthy homes free of indoor air pollutants they are pretty vague about VOCs stating a concern for their "odour." Using materials that smell nice is also a questionable concept. Their free IAQ fact sheet is also very limited in scope. **(63), (64)**

Many of the more dangerous chemicals emitted in buildings are odourless, and it is not obvious from the building biology literature whether they advocate measuring indoor air emissions.

Healthy house builders

There is a small number of house builders and architects claiming to sell, design or build healthy homes. From a careful examination of these, it is clear that these companies tend to cherry-pick different methods and materials, rather than adopting a holistic approach. In some cases, the healthy claim is purely based on the superficial use of low VOC or earth-based paints and finishes and little else that differs from the norm. Others go further selecting timber or fired clay materials rather than concrete. It would be wise to be highly sceptical about current developers claiming to build "healthy homes."

A significant amount of literature about healthy homes has very little to do with the actual materials and indoor air quality in the homes but is instead based on claims about green space provision, play and leisure facilities. A typical example is the Good Homes Alliance though they do at least mention the lack of standards to control indoor pollutants, but without advocating what should be done about it! Their concept of healthy homes is based on the WELL standard. **(65)**

University College London held an event in 2021 entitled "How can we build the healthy home of the future?" As with so many other organisations this turns out to be about "healthy urbanism" and "energy epidemiology," rather than actual healthy homes. **(66)**

The Town and Country Planning Association set up a campaign for healthy homes and sponsored a private members Healthy Homes Bill (discussed in Chapter 4) with limited indoor air quality concerns. It is not surprising that an organisation which has done great work over many years to advocate civilised planning and housing layouts should have perceived healthy homes as about outside space and walkable services, but the 12 principles contained in the Bill are a good start and this may become law. **(67)**

The All Party Parliamentary Group on Healthy Homes and Building, sponsored by BEAMA (ventilation industry), Saint Gobain, (owners of Celotex) and Healthy Developments Ltd, published a white paper which proposes the UK Government needs to establish a cross-departmental committee for health and buildings to champion change, recognising the interaction between buildings, health, education and the economy. While this white paper makes extensive reference to indoor air quality it says little about emissions from construction materials, which are only mentioned once. The paper does draw attention to many important issues and problems but still does not address how to create healthy homes. **(68)**

An organisation called Healthy Homes Solutions has published another white paper, which turns out to have very little to do with healthy homes, other than measures to lower energy bills. There are many organisations that muddle up health with reducing energy use in buildings while ignoring materials and indoor air quality.

> Healthy Homes Solutions (HHS) have researched and identified easy-to install products the public should be encouraged to use to reduce energy consumption, improve heat efficiency and lower bills. Some examples include:
>
> - Digital radiator valves
> - Radiator reflector foil
> - Central heating performance enhancement additive
> - Energy efficient LED light bulbs
> - Water saving products like shower timers and universal plugs
>
> **(69)**

Healthy Homes Solutions provide little information on their website as to who they are, other than providing a postal address in Leigh, Lancashire, which leads us to another company called Healthy Homes Initiatives Ltd. One of the directors has been a director of 35 companies, most of which no longer exist! It was not possible to track down exactly what this company does, but they do give a link to the Energy Saving Trust.

University of Huddersfield has a Healthy Housing Initiative which turns out to be a worthwhile project to assist refugees, though they also post podcasts on cold homes but with little information on what constitutes healthy housing.

> At the HHI, we are committed to advancing the knowledge and understanding of healthy housing, promoting best practices, and advocating for policy change that ensures access to safe and healthy homes for all. Join us in our mission to create a healthier future for all through healthy housing. **(70)**

Ekkist, a small consultancy company, has published a *Healthy Homes Checklist* which they say is a tool made up of 130 practical steps used by some large developers like Quintain and Fore Partnership.

> The Ekkist team reviewed major health-focused building standards, frameworks, guidance documents and scientific research papers relating to health and housing design. After pooling this research, it was reviewed with key industry experts and stakeholders, including Michael Chang from the Office for Health Improvement and Disparities, Spire Building Services, Chapman BDSP, GIA and Sandy Brown. **(71)**

The Checklist, as with so many other initiatives, fails to address how to create healthy homes, referring instead to building "ethos," layout, design and other standard issues but without any reference to indoor air quality or hazardous emissions from building materials; however, Ekkistes do offer WELL certification. **(72), (73)**

Healthy Homes Ireland has produced a report entitled "Our Place: Towards Healthier Greener Homes," published in June 2023. This was an initiative between the Irish Green Building Council and Velux Ltd. Its main recommendations are pretty general but it does propose ensuring that "Indoor Environmental Quality" (IEQ) is fully covered in training for housing professionals and improving regulations to include introduction of VOC labelling for building products. Using the term IEQ generally applies not just to health and wellbeing, but includes temperatures, lighting, air quality and acoustics.

The report contains a number of interesting but confusing and contradictory statements. For instance, on page 16 they say that retrofit can contribute to improvement of IEQ but on page 17 they contradict this.

> "Retrofitting…… can contribute to an improvement of the IEQ of homes" (Page 16)
> "A better insulated and tightly sealed home can worsen the effects of pollutants in the home." (Page 17)

Better than many reports on this topic, they do refer to indoor pollutants and VOC emissions, though without discussing source control measures. **(74)**

Passivhaus

Passivhaus is a design tool for what aspires to be ultra-low energy efficient buildings, which has gained a great deal of popularity with architects and governments. While it is primarily focussed on saving energy, claims are made that it leads to good indoor air quality.

> Passive Houses are designed to ensure that Indoor Air Quality (IAQ) is assured. Fresh air and a condensation-free environment also reduces the risk of mould, damp and cold spots – healthier for the inhabitants, and comfortable. "Passive House buildings are eco-friendly by definition. Passivhaus (also) provides excellent indoor air quality, this is achieved by reducing the air infiltration rates and supplying fresh air which is filtered and post heated by the MVHR unit. **(75)**

Wrapping buildings in plastic and making them extremely airtight, while being dependent on a mechanical ventilation system, is not accepted by everyone as the best route to good indoor air quality. Problems with ducts and a reliance on ventilation are discussed in Chapter 6. Many passivhaus projects continue to rely on petrochemical insulation materials as they believe the questionable claims of higher insulation standards. The Passivhaus Trust states that they are "materials neutral." This means that most passivhaus designers will continue to use high chemical and VOC emitting materials though some are moving toward more natural materials. While good ventilation can improve indoor air quality, this does not mean that sources of indoor air emissions from building materials have been removed. **(76)**

Active House

Active House, or Aktivhaus, was set up in Germany by ecological architects who were concerned at the passivhaus reliance on technocratic solutions. It's now a worldwide movement claiming far more adherents than passivhaus. Active house claims to be more holistic with a stronger emphasis on the health benefits of good daylight and acoustic conditions but is largely unknown in the UK. It's possible for buildings to be certified as complying with Active House principles and for it to be used as a design tool.

> An Active House is a building that offers a healthier and comfortable indoor climate for the occupants without negative impact on the climate – measured in terms of energy, fresh water consumption and the use of sustainable materials. This is a holistic approach to building design that has been adopted in the construction industry and amongst planners and designers. The principles have been tested, and current specifications are based on real data, not just estimates. Setting the user at the centre means to quantify the parameters which matter most to users, with a minimal footprint on the planet. **(77)**, **(78)**

General policies on building construction

There is a perceptible shift in policies in some quarters about the future of building construction, though in general this remains fixated on increasingly air ight buildings with questionable ideas about how they can be more energy efficient. An interesting recent 136-page report by the United Nations represents a significant change in attitudes by referring extensively to the importance of bio-based materials, though focussed almost exclusively on timber and bamboo. Crop-based materials for insulation barely get a mention, for instance.

The report also fails entirely to mention indoor air quality, other than an important reference to the massive use of plastic and a solitary mention of outgassing.

> If managed responsibly, renewable bio-based building materials have a unique capacity to drive reductions in atmospheric carbon by: 1) matching renewable resources to building material applications, at lower carbon footprints, 2) serving as a global carbon sink (see Figure 4.1). Timber is the leading bio-based building material being used at scale.

> Plastics and polymer composites are ubiquitous materials whose use has skyrocketed since the mid-20th century and is projected to more than double by 2060 (OECD 2022b). Plastics are popular due to the low cost and ease of manufacturing. In the United States of America, buildings and construction accounted for 16 per cent of total plastics use in 2015. However, this figure does not account for all the plastics used in the interior furnishings and finishes of buildings, which also can pose risks for the health and well-being of inhabitants from material outgassing. **(79)**

In the UK, on the other hand, the leading figures in the construction field have their heads deeply buried in the sand. The UK *Building the Future Commission*, for instance, is releasing an interim report at a conference held in September 2023 but is worried about a fragile post-pandemic economy and inflationary pressures which have hit budgets on projects, **profit margins** and digital tech investment. Under-investment, squeezed net zero aspirations, politicised housebuilding.

> New safety regulations have caused confusion over how to comply with particular requirements, a lack of joined-up representation for construction inside government and a fragmented skills and education system.

Their focus is largely on digitisation and profitability without any apparent interest in the health effects of buildings and building materials on health and wellbeing. The UK construction industry is in a mess but what is new about that? **(80)**

Not only is the UK construction industry only interested in profits and willing to continue to use chemical and questionable materials and construction systems, failing to learn from past mistakes, even those who talk about healthy homes treat it simply as a slogan or a superficial marketing device, or think it's just about energy efficiency. However, mainstream multi-national building materials companies have recognised the growing importance of natural and bio-based construction materials. Kingspan, one of the leading chemical foam insulation companies, has recently acquired a majority stake Dutch firm Hempflax, one of the biggest producers of hemp-based products, and also in Steico, a German/Polish producer of wood fibre insulation. **(81)**, **(82)**

Currently it is too soon to predict the impact of this development. Other chemical companies have also acquired natural insulation and building material products. Some will welcome this as they assume this will mean greater investment in healthier chemical-free materials, but there is also the danger of diluting or suppressing the market for such materials.

References

(1) Riversong HouseWright. Sheltering within the web of life. Available at: https://riversonghousewright.wordpress.com/about/9-hygro-thermal/ [Accessed 21 Sep. 23].

(2) Twilight Talk February: Unlocking the barriers to the use of bio-based and healthy materials. Available at: www.greenregister.org.uk/events/event/twilight-talk-february-unblocking-the-barriers-to-the-use-of-bio-based-and-healthy-materials-in-retrofitting-and-new-building/ [Accessed 21 Sep. 23].

(3) Harrison, R., Harrison, S., Kimmins, S., and Woolley T. (2000). *Green Building Handbook Volumes 1 and 2*. Abingdon, Oxfordshire: Routledge.

(4) Woolley, T. (2022). *Natural Building Techniques: A Guide to Ecological Methods and Materials*. Marlborough: The Crowood Press.

(5) NHBC Standards (2014). Available at: www.nhbc.co.uk/binaries/content/assets/nhbc/tech-zone/nhbc-standards/nhbc-standards-2014.pdf [Accessed 26 Mar. 24].

(6) Builders Warned of Risks with Natural Materials. (2014). Available at: www.theconstructionindex.co.uk/news/view/builders-warned-of-risk-of-natural-materials [Accessed 21 Sep. 23].

(7) Woolley, T. (2013). *Low Impact Building: Housing Using Renewable Materials*. Hoboken, NJ: Wiley Blackwell, p. 83.

(8) Adamova T. et al. (2020). Volatile Organic Compounds (VOCs) from wood and wood-based panels: methods for evaluation, potential health risks, and mitigation. *Polymers (Basel)*, 12(10): 2289. Available at: www.mdpi.com/2073-4360/12/10/2289 (Accessed 21. Sep 23).

(9) Are Intumescent Paints Toxic? Available at: https://contegointernational.com/intumescent-paints-toxic/ [Accessed 21 Sep. 23].

(10) Sterling OSB Zero Materials Safety Data Revision. (2020). Available at: https://uk.westfraser.com/wp-content/uploads/2019/11/SterlingOSB-Zero-MSDS-July2020.pdf [Accessed 21 Sep. 23].

(11) Norbord Sterling OSB Materials Safety Data Sheet. (2015). Available at: https://dam-assets.apps.travisperkins.group/R0vNP/GPID_1000071569_COSHH_00.pdf [Accessed 21 Sep. 23].

(12) Norbord's Sterling Zero Range-Zero added Formaldehyde (ZAF). Available at: https://uk.westfraser.com/news/norbords-sterlingosb-zero-range-zero-added-formaldehyde-zaf-for-safer-construction-and-healthier-buildings/ [Accessed 21 Sep. 23].

(13) US Environmental Protection Agency. (2011). MDI and Related Compounds Action Plan. Available at: www.epa.gov/sites/default/files/2015-09/documents/mdi.pdf

(14) Gardiner, J. (2023). L&G to Cease Production at Flagship Modular Housing Factory. Available at: www.housingtoday.co.uk/news/landg-to-cease-production-at-flagship-modular-housing-factory/5123028.article [Accessed 21 Sep. 23].

(15) Offsite Concrete Construction TCC/03/64, The Concrete Centre. (2019). www.concretecentre.com/Resources/Publications/Offsite-Concrete-Construction.aspx [Accessed 21 Sep. 23].

(16) Buckton Fields Primary School in Northampton only opened in September 2021 and has been ordered to close by the Department for Education. (2023). Available at: www.pbctoday.co.uk/news/health-safety-news/third-caledonian-modular-school-ordered-to-close-over-structural-safety-concerns/131692/#:~:text=The%20closure%20of%20the%20Northampton,into%20administration%20in%20March%202022 [Accessed 21 Sep. 23].

(17) Lotts, D. (2018). Healthier Homes. Pebble. Available at: https://pebblemag.com/magazine/living/modular-homes-ecofriendly-reasons#:~:text=Healthier%20homes,toxic%20or%20low%20VOC%20materials [Accessed 21 Sep. 23].

(18) Payne S., and Serin B. (2023). The Potential role of Modern Methods of Construction in addressing systemic supply issues UK Collaborative Centre for Housing Evidence (CACHE). https://housingevidence.ac.uk/wp-content/uploads/2023/04/20230310_MMC_PayneSerin_V4.pdf [Accessed 21 Sep. 23].

(19) Weinfass I. (2022). MMC: fire chiefs demand new safety rules www.constructionnews.co.uk/health-and-safety/mmc-fire-chiefs-demand-new-safety-rules-08-12-2022/ [Accessed 21 Sep. 23].

(20) National Fire Chiefs Council Modern Methods of Construction Policy position Statement. (2022). www.nationalfirechiefs.org.uk/write/MediaUploads/Position%20statements/Protection/NFCC_MMC_Policy_Position_Statement__Final.pdf [Accessed 21 Sep. 23].

(21) Fire destroys Shetland's Fair Isle Bird Observatory. (2019). Available at: www.bbc.co.uk/news/uk-scotland-north-east-orkney-shetland-47515175 [Accessed 21 Sep. 23].

(22) Shetland Hotel Destroyed in Overnight Blaze. (2020). Available at: www.bbc.co.uk/news/uk-scotland-north-east-orkney-shetland-53551122 [Accessed 21 Sep. 23].

(23) Fair Isle Bird Observatory Shetland Lighthouse. Available at: www.lighthouse.co.uk/projects/fair-isle-bird-observatory-shetland/ [Accessed 21 Sep. 23].

(24) New Fair Isle Bird Observatory Building Modules Arrive. (2022). Available at: www.hie.co.uk/latest-news/2022/october/19/new-fair-isle-bird-observatory-building-modules-arrive/#:~:text=The%20new%20premises%20will%20include,disastrous%20fire%20in%20March%202019 [Accessed 21 Sep. 23].

(25) Mass Timber Building Science Primer Mass Timber Institute. (2021). Available at: https://academic.daniels.utoronto.ca/masstimberinstitute/building-science-primer/ [Accessed 21 Sep. 23].

(26) Cramer, M. (2023). The Quickest of Overviews of Engineered Wood Products. Available at: https://blogs.napier.ac.uk/cwst/the-quickest-of-overviews-of-engineered-wood-products [Accessed 21 Sep. 23].

(27) Yauk, M. et al. (2020). Evaluating Volatile Organic Compound Emissions from Cross-Laminated Timber Bonded with a Soy-Based Adhesive. *Buildings*, *10*(11), 191. https://doi.org/10.3390/buildings10110191 [Accessed 21 Sep. 23].

(28) Henderson, J., Foster, S., and Bridgestock, M. (2012). Brettstapel What is it? Available at: www.brettstapel.org/Brettstapel/What_is_it.html

(29) Eight Reasons to Use Wood Fibre Insulation. (n.d). Available at: www.backtoearth.co.uk/eight-reasons-to-use-wood-fibre-insulation/#:~:text=Using%20natural%20fibre%20insulation%20products,to%20better%20indoor%20air%20quality [Accessed 21 Sep. 23].

(30) Fuczek, D. et al. (2023). VOC Emissions from Lightweight Wood Fibre Insulation Board. *Forests*, 14(7). Available at: https://www.mdpi.com/1999-4907/14/7/1300 [Accessed 21 Sep. 23].

(31) Clayworks. Available at: https://clay-works.com/ [Accessed 21 Sep. 23].
(32) Clayworks Uses Building Biology Principles. (2023). Available at: https://enkimagazine.com/clayworks-uses-building-biology-principles-healthy-home-design/ [Accessed 21 Sep. 23].
(33) Clayworks Certificates and Verifications. (n.d). Available at: https://clay-works.com/wp-content/uploads/2021/11/clayworks_certificates_and_verifications.pdf [Accessed 21 Sep. 23].
(34) Clay Boards Product Sheet (n.d). Available at: www.claytec.de/Medien%20in%20anderen%20Sprachen/Englisch/09-004_EN.pdf [Accessed 21 Sep. 23].
(35) Hydrated Lime Disinfectants and Algaecides. May 2016 Regulation (EU) No 258/2012.
(36) Available at: https://www.eula.eu/lime-applications/biocides/
(37) EcoRight. (2022). Lime as an Anti-bacterial Coating. Available at: www.ecoright.co.uk/lime-as-an-anti-bacterial-coating/#:~:text=Lime%20Used%20As%20An%20Anti%2DBacterial%20Coating&text=Firstly%2C%20lime%20plaster%20is%20hygroscopic,mould%20growth%20in%20the%20building.
(38) Morrison, A. (n.d). Staying Dry & Healthy: Mold Spores in Straw Bale Homes. Available at: https://strawbale.com/mold-spores-in-straw-bale-homes/ [Accessed 21 Sep. 23].
(39) Sparrow, A. and Stanwix, W. (2014). *The Hempcrete Book*. London: Bloomsbury.
(40) Margent Farm Innovate and Collaborate. (n.d.). Available at: www.margentfarm.com/ [Accessed 21 Sep. 23].
(41) Bio-based Building systems and Products. (n.d.). Available at: https://hemspan.com/ [Accessed 21 Sep. 23].
(42) Wool Protection – Ionic Protect. (n.d). Available at: www.lehner-wool.com/en/brands/isolena.html#:~:text=Ionic%20Protect%C2%AE%20means%20that,function%20as%20an%20insulating%20material [Accessed 21 Sep. 23].
(43) Gil, L. (2009). Cork composites: A review. *Materials* 2(3), 776–789. www.ncbi.nlm.nih.gov/pmc/articles/PMC5445758/ [Accessed 21 Sep. 23].
(44) Sisalwool (n.d.). Available at: https://sisaltech.com/ [Accessed 21 Sep. 23].
(45) What is biocide and why are biocides used in paint products? (2021). Available at: www.paintsforlife.eu/en/blog/what-biocide-and-why-are-biocides-used-paint-products/ [Accessed 21 Sep. 23].
(46) The Volatile Organic Compounds in Paints, Varnishes and Vehicle Refinishing Products Regulations. (2012). Available at: www.legislation.gov.uk/uksi/2012/1715/contents/made [Accessed 21 Sep. 23].
(47) Available at: www.ecocel.ie/wp-content/uploads/2016/09/Ecocel-SDS-safety-data-sheet.pdf.
(48) Warmcel Materials Safety Data Sheet. (2012). Available at: www.kalga.lt/pdf/Warmcel-MSDS-2012.pdf [Accessed 22 Sep. 23].
(49) ASBP Collaborates with Gusto. (2022). Available at: https://asbp.org.uk/asbp-news/anti-greenwash-charter-launched [Accessed 22 Sep. 23].
(50) Charter Signatories. Available at: https://antigreenwashcharter.com/charter-signatories/ [Accessed 22 Sep. 23].
(51) Kellert, S., Heerwagen, J. and Mador, M. (eds) (2008). *Biophilic Design. The Theory, Science and Practice of Bringing Buildings to Life*. Hoboken, NJ: Wiley.
(52) Schneider Guidance on the Green Premium Eco-label. Available at: www.se.com/ww/en/download/document/998-2095-02-25-12AR0_EN/ [Accessed 22 Sep. 23].
(53) Oliver Heath Design BlueAir Event. Available at: https://www.oliverheath.com/case-studies/event-hosting-blue-air/ [Accessed 22 Sep. 23].
(54) Nygaard, A. (2023). Is sustainable certification's ability to combat greenwashing trustworthy? *Frontiers in Sustainability*, 4. Available at: www.frontiersin.org/articles/10.3389/frsus.2023.1188069/full [Accessed 22 Sep. 23].
(55) Fischer, S. The Well Building Standard: Not to be Used Alone. Available at: www.buildinggreen.com/op-ed/well-building-standard-not-be-used-alone [Accessed 22 Sep. 23].
(56) Gaertner, T. Do Health and Wellbeing Standards do Enough. Available at: www.architecture.com/knowledge-and-resources/knowledge-landing-page/do-health-and-wellbeing-standards-do-enough [Accessed 22 Sep. 23].
(57) Living Building Challenge Straw Works. Available at: https://strawworks.co.uk/resources/living-building-challenge/ [Accessed 22 Sep. 23].

(58) Living Building Challenge. Available at: https://living-future.org/lbc/ [Accessed 22 Sep. 23].
(59) Bullitt Center. The Greenest Commercial Building in the World. Available at: https://bullittcenter.org/ [Accessed 22 Sep. 23].
(60) Hanford, J. The Bullitt Center Experience. Available at: www.brikbase.org/sites/default/files/BEST4_3.1%20Hanford.paper_.pdf [Accessed 22 Sep. 23].
(61) International Living Future Institute Section 01 81 14 Construction Indoor Air Quality. Available at: https://living-future.org/wp-content/uploads/2022/05/Living-Building-Challenge-Sample-018114-Construction-IAQ.pdf [Accessed 22 Sep. 23].
(62) Various Authors. (2023). *The Regenerative Materials Movement: Dispatches from Practitioners, Researchers, and Advocates*. Bainbridge Island, WA: Ecotone Publishing.
(63) Building Biology Association UK. Available at: https://buildingbiology.co.uk/ [Accessed 22 Sep. 23].
(64) IAQ for You Factsheet. Available at: https://buildingbiologyinstitute.org/free-fact-sheets/indoor-air-quality-ieq-for-you/
(65) Good Homes Alliance Healthy Homes. Available at: https://goodhomes.org.uk/what-we-do/healthy-homes/ [Accessed 22 Sep. 23].
(66) Available at: www.ucl.ac.uk/health-of-public/events/2021/jul/how-can-we-build-healthy-home-future [Accessed 22 Sep. 23].
(67) TCPA Campaign for Healthy Homes. Available at: www.tcpa.org.uk/collection/campaign-for-healthy-homes/ [Accessed 22 Sep. 23].
(68) Building Our Future. Available at: https://healthyhomesbuildings.org.uk/wp-content/uploads/2018/10/HHB-APPG-White-Paper-V1.pdf [Accessed 22 Sep. 23].
(69) Healthy Homes Solutions. Available at: https://healthyhomessolutions.co.uk/ [Accessed 22 Sep. 23].
(70) Welcome to the Healthy Housing Initiative. Available at: http://healthy-housing.uk/#:~:text=At%20the%20HHI%2C%20we%20are,for%20all%20through%20healthy%20housing [Accessed 22 Sep. 23].
(71) The Ekkist Healthy Homes Checklist. Available at: www.ekkist.co/healthyhomeschecklist [Accessed 24 Mar. 24].
(72) Ekkist launches Healthy Homes Checklist to help tackle UK's Unhealthy Housing. Available at: www.showhouse.co.uk/news/ekkist-launches-healthy-homes-checklist-to-tackle-unhealthy-housing-crisis/ [Accessed 22 Sep. 23].
(73) The Ekkist Healthy Homes Checklist. Available at: https://www.ekkist.co/healthyhomeschecklist/ [Accessed 22 Sep. 23].
(74) Launch of Healthy Homes Ireland Recommendations. Available at: www.igbc.ie/events/launch-of-healthy-homes-ireland-recommendations/ [Accessed 22 Sep. 23].
(75) Norrsken Are Passive Houses Healthy? Available at: www.norrsken.co.uk/newsitem/passive-houses-benefits#:~:text=Passive%20Houses%20are%20designed%20to,for%20the%20inhabitants%2C%20and%20comfortable [Accessed 22 Sep. 23].
(76) The Healthy Homes Ltd. Available at: www.thehealthyhome.uk/passivhaus [Accessed 22 Sep. 23].
(77) The author is a board member of Active House UK.
(78) Healthy Buildings for People and Planet. Available at: www.activehouse.info/wp-content/uploads/2022/02/Healthy-Buildings-for-People-Planet.pdf [Accessed 22 Sep. 23].
(79) Building Materials and the Climate: Constructing a New Future 2023. United Nations Environment Programme Job number: DTI/2563/PA. Available at: www.unep.org/resources/report/building-materials-and-climate-constructing-new-future [Accessed 22 Sep. 23].
(80) The Building the Future Commission: The Interim Findings Revealed. Available at: www.bdonline.co.uk/briefing/the-building-the-future-commission-the-interim-findings-revealed/5125181.article [Accessed 22 Sep. 23].
(81) Kingspan insulation has signed a deal for the acquisition of a majority stake in HempFlax building solutions gmbh. Available at: www.thermo-hanf.de/en/about/company/ [Accessed 3 October 23].
(82) Kingspan announces majority stake in Steico. Available at: www.kingspangroup.com/en/news-insights/kingspan-announces-acquisition-of-majority-stake-in-steico-se/ [Accessed 3 October 23].

9 Afterthoughts

> Is the desire to create a safer better world for all of us really that 'radical'? Isn't it something we should all want not just for ourselves but for the generations that will come after us?
>
> (Mikaela Loach) **(1)**

While doing the research and writing for this book, the issue kept resurfacing as to why building emission deniers exist. Are they better scientists than this author and how have they convinced themselves that chemicals in building materials is not something we should worry about? Some say there is little evidence of harm from chemicals in building materials, and that problems with traffic pollution, middle class lifestyle, wood-burning stoves and cooking on gas are far more serious.

One explanation is that it is relatively easy to criticise middle class people making "lifestyle choices" of installing wood burning stoves, whereas taking on the massive petrochemical and multi-national construction materials sector is just a bit harder. Scientific conclusions reflect the bias of scientists and many are not willing to take the apparently radical step of challenging the status quo. The bias in this book on the other hand should be only too obvious in the opposite direction. For many, showing concern about chemicals in buildings is something they are happy to turn a blind eye to. Without obvious causal evidence, it's easier just to deny there is a problem, but this represents bias. Adopting the precautionary principle, on the other hand, means that rejecting potentially health harming materials involves searching for healthier and better alternatives.

Fortunately, the alternatives are there and it is possible to design and build a house largely free of chemicals and plastic. Not only will this be a healthier house but it can be a better, warmer and more robust house. To do this, however, involves extra work and a commitment to search out and source the healthier materials and find builders willing to use them. It may involve battling with the authorities who don't want to approve a healthier and better solution as the whole system is geared toward doing what everyone else does! There is so much pressure to stick to conventional solutions that often clients will lose their nerve.

Are architects, building control officials, academics, scientists, civil servants and NGOs in the UK in league with the chemical companies? This seems unlikely, but many scientists and others tend to support the status quo and resist what they perceive as radical ideas. Political thinking is generally very short term and resistant to change and this impacts on science.

> Evidence-based policymaking is an attractive paradigm, but in a parliamentary democracy, political considerations will almost always take precedence. Policy decision-making is bound by what is politically feasible. If a researcher's work points to policy choices beyond

DOI: 10.1201/9781003129226-9

> these boundaries, it is usually unrealistic to expect policy change, however compelling the research findings. And if bits of an academic's research help to justify a politician's ideology or perspective, it is usually the findings that will be most likely to result in policy change. **(2)**

> Observers often assume that when people are dissatisfied, they will demand changes. But cognitive and behavioural scientists know that frequently is not the case, because a situation called "the default effect" prevails. **(3)**

There is a tendency to default to accepting the status quo and resisting more radical calls for change or to question current thinking. It has become apparent when researching this book that there is a passive consensus within the indoor air quality world that has pushed the issue of chemical emissions from building materials into the background. There is also a problem of closed minds, evidenced by the unwillingness of expert groups to search the literature properly or to be highly selective when writing reports about indoor air quality. This author can certainly be accused of selectivity as I have searched for the literature that supports the argument that we should be worried about chemicals in building materials.

A new book by Jay Owens called "Dust," previewed in "the long read" in the *Guardian,* shows a surprising awareness of indoor air quality and health issues. Clearly not a fan of electric cars, which Owens points out generates more tyre dusts and road wear because of the excessive weight of electric cars, she is also aware of the dangers of endocrine-disrupting flame retardants which "Can cause cancer, decrease fertility, impact cognitive ability and cause thyroid disease." Owens has not picked this up from the DEFRA AQEG and NICE reports, so it is encouraging that a journalist non-expert has been so perceptive in their awareness of indoor air quality! **(4)**

The culture of technocracy

It is possible that there is a cultural problem in the UK, that respect for industry and technocrats overwhelms rational thought. It would be easy to blame Thatcher for the cultural revolution she introduced, but before her Harold Wilson and his infamous White Heat of Technology speech, which led to the disastrous system-built tower blocks, indicates that this is a problem which crosses political boundaries. As a young student, I spent a few months on a summer job in the Edinburgh office of the National Building Agency, at a time in 1967 when Government was endorsing and approving prefabricated and other concrete building systems. It was hard to work out what many of the staff in the office were doing with their time, but there was one less senior member of staff in the corner of the office who wielded a big rubberstamp marked APPROVED. Every day couriers would arrive with big roles of blueprints (this was before computer-aided design) and they would be rolled out on his desk, and he worked away stamping all the drawings APPROVED! I dared to ask what assessment process they were going through as he pointed to his rubberstamp and big ink pad. I asked if he had a similar stamp saying REJECTED and he looked puzzled. The rest, of course, was history, as millions of pounds of concrete and asbestos was blown up within a decade or so. Yorkshire Development Group (YDG), Taylor Woodrow's Ronan Point, Red Road Flats and many, many, more leading tragically to Grenfell.

There seems to be a built-in willingness on the part of many professionals to accept certain kinds of innovative solutions if they are backed by big finance and reject others that seem too "alternative." As pointed out earlier in the book untried and high-risk technology such as flammable foam insulation is accepted by architects as though it has been around for decades while rejecting "innovative" hempcrete which has been in use for just as long.

The cladding crisis following Grenfell has slipped from the news but many thousands of people are still living in apartment buildings where the cladding has been removed but not replaced. Thousands of people cannot sell their flats, and many have been bankrupted by spiralling insurance and services charges. Government bail-out schemes have barely touched the problem and the Government has failed to pursue the builders, developers and materials suppliers who were responsible for paying compensation. There is no space here to document the hundreds of examples of this problem but Brocklehurst Court in Macclesfield, a modest three-storey block, built with perfectly safe brick walls but then covered in a rendered polystyrene EWI scheme is an interesting example.

> It was only in September last year (2022) that the residents of Brocklehurst Court in Macclesfield, Cheshire, found that they had been caught up in the cladding crisis. When Jones Associates, the managing agent of the property, went to renew its insurance on the building, they saw premiums shoot up from £2,768 to £24,450 a year – a whopping 783% increase. That surge is due to the expanded polystyrene (EPS) cladding system on Brocklehurst Court, which was put on the building in 2013 using a government grant. Residents insist the fire service has told them the cladding system is safe. **(5)**

Whether or not external rendered polystyrene is fire safe is a matter of debate, but the insurers clearly perceive a risk, even if the local fire brigade does not. In reality the risk is probably fairly low but what is at the heart of this story is the willingness of Government grant schemes to approve something that has not been properly tried and tested… almost certainly because a big commercial company has met them at a seminar or trade show and told them that its ok. Following thousands of retrofit disasters, the response is to introduce more bureaucracy such as PAS 2035, "Trust a Trader," and training schemes creating new bureaucratic roles such as retrofit designers and co-ordinators, but they don't discuss which are the safest materials to use.

This is the same culture that led to the recent aerated autoclave concrete collapses in schools and other public buildings. How is it possible that in the pursuit of cheapness, poor quality innovative materials are so readily accepted? Not only are hundreds of buildings on the edge of collapse, but it has emerged that many aerated concrete buildings are also contaminated with asbestos. **(6)**

While children and teachers suffer from the problem of closed buildings, it is claimed that the Department of Education has been spending £111 million on luxury refurbishment of its own offices in London. **(7)**

This is not just a UK problem but similar problems can be found in Ireland. In counties Donegal and Mayo in the Republic of Ireland hundreds (possibly thousands) of houses were built with a concrete using "mica" where the houses are now crumbling. The Irish Government approved a €20 million repair scheme, though the problem is still a long way from resolution. In County Clare, a similar series of disasters occurred where "pyrites" were used in concrete blocks. It has only taken 20 years for these construction systems to collapse. **(8), (9)**

Ireland has not had a Grenfell but there have been many smaller scandals with bad and unsafe building practices, particularly during the Celtic Tiger period, well documented by O'Broin who exposed the fire safety hazards as well as structural defects. **(10)**

Meanwhile dozens of companies, hoping to make a killing through Government-promoted Modern Methods of Construction (MMC), have gone bankrupt as discussed earlier in the book. Serious fire risk concerns have led to panic-stricken measures such as highly expensive sprinkler systems in individual two-storey houses in Wales as well as tower blocks.

New regulations and regulators have been brought in following Grenfell, but with a clear sense of shutting the barn door too late. The Government and energy agencies continue to promote unsuitable cavity wall and other insulation schemes, ruining the homes and often the lives of thousands of occupants, with a guarantee scheme that often fails to pay out. Bureaucratic red tape retrofit procedures, dressed up as standards, and use of terms like Academy and Hub, which fail to recognise that sealing up buildings with cement and plastic leads to mould and damp and even deaths of children. Government advisors who seem more concerned with the profits of industry than the genuine safety of the public, and advisory committees that are dominated by commercially vested interests. It all adds up to a pattern that means that the use of chemicals, the emissions they generate and the health dangers that result are barely on the minds of the policymakers. And this argument doesn't even touch on lies about diesel emissions by car makers and the billions made in Covid profits by private companies, linked to politicians, that ended up in waste mountains in the New Forest. **(11)** When things go wrong or injustices occur it seems to take generations for inquiries to be held and compensation to be paid whether it was the Hillsborough tragedy or the Contaminated Blood Scandal and the shocking legacy of the nuclear test veterans. **(12)**

People suffering from damp and mould

As the writing of this book was nearing completion it was hard to avoid being distracted by weekly requests to visit and investigate mould and damp houses. Other professionals, journalists and ordinary people would somehow find a way to get in touch despite the author not advertising his services as there did not seem to be anyone else to turn to. Academic and voluntary sector events about fuel poverty and housing conditions seem well funded but it is just talk and no action. Healthy Homes and Beyond in Berwick in Northumberland offers a service throughout the UK investigating cases of mould and damp and indoor air quality but is just run by two people. Why are there not similar organisations throughout the land? **(13)**

Are there dangers in greenwash solutions?

While it is possible to build or renovate with chemical-free "healthy" materials, it will be important to watch out for snake-oil solutions offered by greenwash and well-wash companies. While they may claim to be producing healthy and low-carbon buildings, some of the innovative "green and healthy" building systems emerging into the market sometimes involve compromised and poor quality materials. New systems are being promoted using composite timber boards with a variety of plastic and synthetic materials. An English company says it is "a pioneering collective with a mission to make out built environment more innovative and sustainable through digital manufacturing technologies." Various "click and connect" off-site construction systems, some using 3-D printing, many using OSB timber cassettes, wrapped in plastic and approved by the Build Off-Site Property Assurance Scheme (BOPAS) are emerging. Others use off-site constriction with Structural Insulation Panels (SIP) made with PET recycled plastic (polyethylene terephthalate) When contacted these companies are sometimes evasive about the materials they use, particularly flame retardants (commercially confidential), and refuse to discuss issues like indoor air quality. It's hard to find health and safety data sheets and attempts to visit some highly publicised buildings have been unwelcome. These could be the potential disasters of the future, as we continue to have an approval system in the UK where high-pressure chemical industry backed salesmanship holds sway and building regulations continue to be inadequate.

Government agencies continue to promote "low energy" plastic construction, often with subsidies. There is a handful of demonstration "Active Houses" which are covered with high fire risk, stick-on photovoltaic cells covering walls and roofs, backed by foam insulation, described as buildings as power stations. One scheme of 16 2/3 bedroom houses and apartments in Wales is reported to have cost at least 12% more than conventional housing, and some reports suggest much more. **(14)**

Some of the better innovative healthy solutions are also problematic as they are under pressure from two sides. Mainstream multi-national cement and plastic companies are trying to take over bio-based materials production from one side. From the other side are bandwagon jumper amateur enthusiasts who think that anyone with limited skills and experience can do eco-construction, having viewed a few internet sites. Somewhere in the middle are plenty of excellent and competent people but struggling to cope with a lack of certification and approvals, yet able to build good buildings. It is possible to build healthy buildings with minimal chemicals but consumer demand will need to be stronger before this becomes commonplace.

Positive examples

Small companies like Wellspring homes in Wales are emerging that claim "a true commitment to sustainability, and delivering a healthier, enriched way of life for all of its residents." Building with hempcrete and other natural materials they are able to deliver affordable, low-carbon and excellent indoor air quality solutions. However, social housing agencies such as housing associations, council and private developers refuse to consider such materials and continue to use chemical synthetic plastic approaches. **(15)**

There are plenty of one-off healthy eco homes being built, but these are frequently dismissed as for self-indulgent wealthy middle class people building in the countryside. Moving eco-healthy building into the mainstream seems a bridge too far at present. As poorer economies in the global south catch up with the developed north, many people want concrete and plastic buildings as these are seen as a symbol of wealth and success. As global warming continues and temperatures rise, gas guzzling mechanical ventilation and air conditioning are seen as the only solutions rather than using low-energy, passive, natural solutions.

Coping with the animosity!

There will be a lot of hostility to this book and the implications of reducing chemical use will not be welcomed by building materials companies, academic and scientist fellow travellers and professionals in the construction industry. Business as usual is their approach, and they will double their efforts behind the scenes to come up with diversion tactics like getting us to stop cooking with gas, even though the current Government looks likely to delay the end of gas heating. The cases made in this book are not perfect and they suffer from a continuing lack of evidence and proper research but hopefully some people will see that there are real issues here. I will continue to battle on behalf of those people who suffer from the consequences of chemicals in bad buildings.

> No, I'll stand my ground
> Won't be turned around
> And I'll keep this world from draggin' me down
> Gonna stand my ground
> And I won't back down
>
> (Tom Petty) **(16)**

References

(1) Loach, M. (2023). *It's not that Radical: Climate Action to Transform Our World*. London: Dorling Kindersley.
(2) Lloyd, J. (2023). Should academics be expected to change policy? Available at: https://blogs.lse.ac.uk/politicsandpolicy/should-academics-be-expected-to-change-policy/ [Accessed 20 Sep. 23].
(3) Scholars Strategy. (2017). Network. Available at: https://scholars.org/contribution/why-dissatisfied-people-settle-status-quo [Accessed 20 Sep. 23].
(4) Owens, J. (2023). *Dust: The Modern World in a Trillion Particles*. London: Hodder and Stoughton.
(5) Geraghty, L. (2023). www.bigissue.com/news/housing/beyond-grenfell-the-cladding-crisis-still-gripping-the-uk/ [Accessed 20 Sep. 23].
(6) Murray, J. (2023). Available at: www.theguardian.com/education/2023/sep/03/how-asbestos-could-slow-efforts-fix-crumbling-concrete-english-schools [Accessed 20 Sep. 23].
(7) Whittaker, F. (2023). Available at: https://schoolsweek.co.uk/dfe-spent-111m-on-offices-it-wont-inhabit/ [Accessed 20 Sep. 23].
(8) Wilson, D. (2021). Donegal: Crumbling house homeowners want new support scheme. www.bbc.co.uk/news/world-europe-57204910.amp [Accessed 20 Sep. 23].
(9) Tighter Regulation needed to ensure pyrite scandal not repeated. (2022). https://clarechampion.ie/clare-pyrite-scandal/ [Accessed 20 Sep. 23].
(10) O'Broin, E. (2021). *Defects Living with the Legacy of the Celtic Tiger*. Kildare, Ireland: Merrion Press.
(11) Somerville, E. (2023). Mountain of Unused PPE dumped in nature reserve. June 18 2023. Available at: www.telegraph.co.uk/news/2023/06/18/ppe-covid-mountain-new-forest-hampshire-abandoned/ [Accessed 20 Sep. 23].
(12) Legacy of the atomic bomb recognition for the atomic test survivors. Available at: www.labrats.international/family.
(13) Welcome to Healthy Homes and Beyond. Available at: https://Healthyhomesandbeyondiaq.org.uk.
(14) Bradfield, E. Tenants to move into new 'active homes' in Neath. Available at: www.walesonline.co.uk/news/local-news/tenants-move-new-active-homes-16905851.
(15) Wellspring Curating Communities restoring the earth. Available at: https://wellspringhomes.co.uk/developments.
(16) Petty T. I Won't Back Down. (1989). Track 2 on Full Moon Fever. Available at: https://genius.com/Tom-petty-i-wont-back-down-lyrics.

Appendix

Agreed EU-LCI values December 2021

No.	*CAS no.*	*Compound*	*EU-LCI (µg/m³)*	*Status of EU-LCI value*	*Year of adoption*
1		***Aromatic hydrocarbons***			
1-1	108-88-3	**Toluene**	**2900**	**Derived EU-LCI**	**2013**
1-2	100-41-4	**Ethylbenzene**	**850**	**Derived EU-LCI**	**2013**
1-3	1330-20-7 106-42-3 108-38-3 95-47-6	**Xylene (o-, m-, p-) and mix of o-, m- and p-xylene isomers**	**500**	**Derived EU-LCI**	**2013**
1-4	98-82-8	**Isopropylbenzene (cumene)**	**1700**	**Derived EU-LCI**	**2017**
1-5	103-65-1	**n-Propylbenzene**	**950**	**Derived EU-LCI (read-across)**	**2013**
1-6	108-67-8 95-63-6 526-73-8	**Trimethylbenzene (1,2,3-,1,2,4-,1,3,5-)**	**450**	**Derived EU-LCI**	**2013**
1-7	611-14-3	**2-Ethyltoluene**	**550**	**Derived EU-LCI (read-across)**	**2014**
1-8	527-84-4 535-77-3 99-87-6 25155-15-1	**Cymene (o-, m-,p-,) (1-isopropyl-2(3,4)- methylbenzene) and mix of o-, m-, and p- cymene**	**1000**	**Ascribed EU-LCI**	**2013**
1-9	95-93-2	**1,2,4,5-Tetramethylbenzene**	**250**	**Derived EU-LCI (read-across)**	**2016**
1-10	104-51-8	**n-Butylbenzene**	**1100**	**Derived EU-LCI (read-across)**	**2014**
1-11	99-62-7 100-18-5	**Diisopropylbenzene (1,3-, 1,4-)**	**750**	**Derived EU-LCI (read-across)**	**2013**
1-12	2189-60-8	**Phenyl octane and isomers**	**1100**	**Derived EU-LCI (read-across)**	**2013**
1-16	100-42-5	**Styrene**	**250**	**Derived EU-LCI**	**2013**
1-17	98-83-9	**2-Phenylpropene (α-methylstyrene)**	**1200**	**Derived EU-LCI**	**2018**
1-18	637-50-3	**1-Propenyl benzene (ß-methyl styrene)**	**1200**	**Derived EU-LCI (read-across)**	**2019**
1-20	611-15-4 100-80-1 622-97-9 25013-15-4	**Vinyl toluene (o-, m-, p-) and mix of o-, m-, and p-vinyl toluene**	**1200**	**Derived EU-LCI**	**2018**
1-23	91-20-3	**Naphthalene**	**10**	**Derived EU-LCI**	**2015**
1-24	91-17-8	**Decahydronaphthalene**	**200**	**Derived EU-LCI**	**2019**
1-25	95-13-6	**Indene**	**450**	**Ascribed EU-LCI**	**2013**

No.	CAS no.	Compound	EU-LCI (μg/m³)	Status of EU-LCI value	Year of adoption
2		***Saturated aliphatic hydrocarbons (n-, iso- and cyclo-)***			
2-1	110-54-3	**n-Hexane**	**4300**	**Derived EU-LCI**	**2016**
2-2	110-82-7	**Cyclohexane**	**6000**	**Ascribed EU-LCI**	**2013**
2-3	108-87-2	**Methyl cyclohexane**	**8100**	**Ascribed EU-LCI**	**2013**
2-4	142-82-5	**n-Heptane**	**15000**	**Derived EU-LCI**	**2018**
2-5		**Other saturated aliphatic hydrocarbons C6-C8**	**14000**	**Derived EU-LCI (read-across)**	**2018**
2-6		**Other saturated aliphatic hydrocarbons C9-C16**	**6000**	**Ascribed EU-LCI**	**2013**
3		***Terpenes***			
3-1	498-15-7	**3-Carene**	**1500**	**Ascribed EU-LCI**	**2013**
3-2	80-56-8	**α-Pinene**	**2500**	**Derived EU-LCI**	**2013**
3-3	127-91-3	**ß-Pinene**	**1400**	**Ascribed EU-LCI**	**2013**
3-4	138-86-3 5989-27-5 5989-54-8	**Limonene**	**5000**	**Derived EU-LCI**	**2014**
3-5		**Other terpene hydrocarbons**	**1400**	**Ascribed EU-LCI**	**2013**
4		***Aliphatic alcohols***			
4-1	75-65-0	**2-Methyl-2-propanol (tert-butanol)**	**620**	**Ascribed EU-LCI**	**2013**
4-2	78-83-1	**2-Methyl-1-propanol**	**11000**	**Derived EU-LCI**	**2016**
4-3	71-36-3	**1-Butanol**	**3000**	**Ascribed EU-LCI**	**2013**
4-4	71-41-0 30899-19-5 94624-12-1 6032-29-7 584-02-1 137-32-6 123-51-3 598-75-4 75-85-4 75-84-3	**1-Pentanol (all isomers)**	**730**	**Ascribed EU-LCI**	**2013**
4-5	111-27-3	**1-Hexanol**	**2100**	**Ascribed EU-LCI**	**2013**
4-6	108-93-0	**Cyclohexanol**	**2000**	**Ascribed EU-LCI**	**2013**
4-7	104-76-7	**2-Ethyl-1-hexanol**	**300**	**Derived EU-LCI**	**2014**
4-8	111-87-5	**1-Octanol**	**1700**	**Derived EU-LCI**	**2016**
4-9	123-42-2	**4-Hydroxy-4-methyl-pentane-2-on (diacetone alcohol)**	**960**	**Ascribed EU-LCI**	**2013**
4-11*	105-08-8	**1,4-Cyclohexanedimethanol**	**8300**	**Derived EU-LCI**	**2021**
5		***Aromatic alcohols***			
5-1	108-95-2	**Phenol**	**70**	**Derived EU-LCI**	**2017**
5-2	128-37-0	**BHT (2,6-di-tert-butyl-4-methylphenol)**	**100**	**Ascribed EU-LCI**	**2013**
5-3	100-51-6	**Benzyl alcohol**	**440**	**Ascribed EU-LCI**	**2013**
6		***Glycols, glycol ethers, glycol esters***			
6-1	107-21-1	**Ethandiol (ethylenglykol)**	**3400**	**Derived EU-LCI**	**2016**
6-2	96-49-1	**Ethylene carbonate**	**4800**	**Derived EU-LCI (read-across)**	**2020**
6-3	7397-62-8	**Butyl glycolate**	**900**	**Derived EU-LCI**	**2019**
6-4	111-46-6	**Diethylene glycol**	**5700**	**Derived EU-LCI (read-across)**	**2016**

(*Continued*)

No.	*CAS no.*	*Compound*	*EU-LCI ($\mu g/m^3$)*	*Status of EU-LCI value*	*Year of adoption*
6-5	57-55-6	**Propylene glycol (1,2-dihydroxypropane)**	**2100**	**Derived EU-LCI**	**2016**
6-6*	108-32-7	**Propylene carbonate**	**1800**	**Derived EU-LCI**	**2021**
6-7	623-84-7	**Propylene glycol diacetate**	**1600**	**Derived EU-LCI (read-across)**	**2018**
6-8	110-98-5 25265-71-8	**Dipropylene glycol**	**670**	**Ascribed EU-LCI**	**2013**
6-9	110-63-4	**1,4-Butanediol**	**2000**	**Ascribed EU-LCI**	**2013**
6-10	107-41-5	**Hexylene glycol (2-methyl-2,4- pentanediol)**	**3500**	**Derived EU-LCI**	**2018**
6-11	6846-50-0	**2,2,4-Trimethylpentanediol diisobutyrate**	**1300**	**Derived EU-LCI**	**2018**
6-12	109-86-4	**Ethylene glycol monomethyl ether (2-methoxyethanol)**	**100**	**Derived EU-LCI**	**2018**
6-13	110-49-6	**2-Methoxyethyl acetate**	**150**	**Derived EU-LCI (read-across)**	**2018**
6-14	110-71-4	**1,2-Dimethoxyethane**	**100**	**Derived EU-LCI**	**2020**
6-15	111-96-6	**Diethylene glycol dimethyl ether (1-methoxy-2-(2-methoxy-ethoxy)-ethane)**	**28**	**Ascribed EU-LCI**	**2013**
6-16	25265-77-4	**2,2,4-Trimethyl-1,3-pentanediol monoisobutyrate**	**850**	**Derived EU-LCI**	**2018**
6-17	109-59-1	**Ethylene glycol isopropylether (2-methylethoxyethanol)**	**220**	**Ascribed EU-LCI**	**2013**
6-18	112-49-2	**Triethylene glycol-dimethyl ether**	**150**	**Derived EU-LCI**	**2019**
6-19	110-80-5	**Ethylene glycol monoethyl ether (2-ethoxyethanol)**	**600**	**Derived EU-LCI**	**2016**
6-20	111-15-9	**2-Ethoxyethyl acetate**	**900**	**Derived EU-LCI (read-across)**	**2016**
6-21	629-14-1	**1,2-Diethoxyethane**	**150**	**Derived EU-LCI**	**2020**
6-22	111-90-0	**Diethylene glycol monoethyl ether (2-(2-ethoxyethoxy)ethanol)**	**350**	**Ascribed EU-LCI**	**2013**
6-23	2807-30-9	**Ethylene glycol monoisopropyl ether (2-propoxyethanol)**	**860**	**Ascribed EU-LCI**	**2013**
6-24	111-76-2	**Ethylene glycol monobutylether (2-butoxyethanol)**	**1600**	**Derived EU-LCI**	**2016**
6-25	112-07-2	**2-Butoxyethyl acetate**	**2200**	**Derived EU-LCI (read-across)**	**2016**
6-26	112-34-5	**Diethylene glycol monobutylether**	**350**	**Derived EU-LCI**	**2019**
6-27	124-17-4	**Diethylene glycol monomethyl ether acetate (butyldiglykolacetate, 2-(2-butoxyethoxy) ethyl acetate)**	**850**	**Ascribed EU-LCI**	**2013**
6-28	122-99-6	**2-Phenoxyethanol**	**60**	**Derived EU-LCI**	**2016**
6-29	112-25-4	**Ethylene glycol n-hexyl ether (2-hexoxyethanol)**	**900**	**Derived EU-LCI**	**2019**
6-30	112-59-4	**Diethylene glycol n-hexyl ether (2-(2- hexoxyethoxy)-ethanol)**	**400**	**Derived EU-LCI (read-across)**	**2019**
6-31	107-98-2	**Propylene glycol monomethyl ether (1-methoxy-2-propanol)**	**7900**	**Derived EU-LCI**	**2018**
6-32	1589-47-5	**1-Propylene glycol 2-methyl ether (2-methoxy-1-propanol)**	**19**	**Ascribed EU-LCI**	**2013**
6-33	70657-70-4	**1-Propylene glycol 2-methyl ether acetate (2-methoxy-1-propyl acetate)**	**28**	**Ascribed EU-LCI**	**2013**

No.	*CAS no.*	*Compound*	*EU-LCI ($\mu g/m^3$)*	*Status of EU-LCI value*	*Year of adoption*
6-34	7778-85-0	**1,2-Propylene glycol dimethyl ether**	**500**	**Derived EU-LCI**	**2019**
6-35	34590-94-8	**Dipropylene glycol monomethyl ether**	**3100**	**Ascribed EU-LCI**	**2013**
6-36	88917-22-0	**Dipropylene glycol monomethyl ether acetate**	**950**	**Derived EU-LCI (read-across)**	**2019**
6-37	29911-27-1 29911-28-2	**Dipropylene glycol mono-n-propylether**	**200**	**Derived EU-LCI (read-across)**	**2019**
6-38	35884-42-5 132739-31-2	**Dipropylene glycol mono-n(t)-butylether**	**250**	**Derived EU-LCI**	**2019**
6-39	20324-33-8 25498-49-1	**Tripropylene glycol mono-methylether**	**1200**	**Derived EU-LCI**	**2018**
6-40	63019-84-1 89399-28-0 111109-77-4	**Dipropylene glycol dimethyl ether**	**1300**	**Ascribed EU-LCI**	**2013**
6-41*	2517-43-3	**3-Methoxy-1-butanol**	**1700**	**Derived EU-LCI**	**2021**
6-42*	1569-01-3 30136-13-1	**1,2-Propylene glycol n-propylether**	**5200**	**Derived EU-LCI**	**2021**
6-43	5131-66-8 29387-86-8 15821-83-7 63716-40-5	**1,2-Propylene glycol n-butylether**	**650**	**Derived EU-LCI**	**2018**
6-44	104-68-7	**Diethylene glycol phenylether**	**80**	**Derived EU-LCI (read-across)**	**2019**
6-45	126-30-7	**Neopentyl glycol**	**8700**	**Derived EU-LCI**	**2020**
7		***Aldehydes***			
7-1	50-00-0	**Formaldehyde**	**100**	**Derived EU-LCI**	**2016**
7-2	75-07-0	**Acetaldehyde**	**300**	**Derived EU-LCI**	**2020**
7-3	123-38-6	**Propanal**	**650**	**Derived EU-LCI**	**2018**
7-4	123-72-8	**Butanal**	**650**	**Derived EU-LCI**	**2013**
7-5	110-62-3	**Pentanal**	**800**	**Derived EU-LCI (read-across)**	**2013**
7-6	66-25-1	**Hexanal**	**900**	**Derived EU-LCI (read-across)**	**2013**
7-7	111-71-7	**Heptanal**	**900**	**Derived EU-LCI (read-across)**	**2013**
7-8	123-05-7	**2-Ethyl-hexanal**	**900**	**Derived EU-LCI (read-across)**	**2013**
7-9	124-13-0	**Octanal**	**900**	**Derived EU-LCI (read-across)**	**2013**
7-10	124-19-6	**Nonanal**	**900**	**Derived EU-LCI (read-across)**	**2013**
7-11	112-31-2	**Decanal**	**900**	**Derived EU-LCI (read-across)**	**2013**
7-12	4170-30-3 123-73-9 15798-64-8	**2-Butenal (crotonaldehyd)**	**5**	**Derived EU-LCI**	**2015**
7-13	1576-87-0 764-39-6 31424-04-1	**2-Pentenal**	**7**	**Derived EU-LCI (read-across)**	**2015**

(*Continued*)

No.	*CAS no.*	*Compound*	*EU-LCI ($\mu g/m^3$)*	*Status of EU-LCI value*	*Year of adoption*
7-14	6728-26-3 505-57-7 16635-54-4 1335-39-3 73543-95-0	**Hexenal**	**7**	**Derived EU-LCI (read-across)**	**2015**
7-15	2463-63-0 18829-55-5 57266-86-1 29381-66-6	**2-Heptenal**	**7**	**Derived EU-LCI (read-across)**	**2015**
7-16	2363-89-5 2548-87-0 25447-69-2 20664-46-4	**2-Octenal**	**7**	**Derived EU-LCI (read-across)**	**2015**
7-17	2463-53-8 18829-56-6 60784-31-8	**2-Nonenal**	**7**	**Derived EU-LCI (read-across)**	**2015**
7-18	3913-71-1 2497-25-8 3913-81-3	**2-Decenal**	**7**	**Derived EU-LCI (read-across)**	**2015**
7-19	2463-77-6 53448-07-0 1337-83-3	**2-Undecenal**	**7**	**Derived EU-LCI (read-across)**	**2015**
7-20	98-01-1	**Furfural**	**10**	**Derived EU-LCI**	**2017**
7-21	111-30-8	**Glutaraldehyde**	**1**	**Derived EU-LCI**	**2018**
8		***Ketones***			
8-1	78-93-3	**2-Butanone (ethylmethylketone)**	**20000**	**Derived EU-LCI**	**2016**
8-2	563-80-4	**3-Methyl-2-butanone**	**7000**	**Ascribed EU-LCI**	**2013**
8-3	108-10-1	**4-Methyl-2-pentanone (methylisobutylketone)**	**1000**	**Derived EU-LCI**	**2016**
8-4	120-92-3	**Cyclopentanone**	**1200**	**Derived EU-LCI**	**2020**
8-5*	108-94-1	**Cyclohexanone**	**1400**	**Derived EU-LCI**	**2021**
8-6	1120-72-5	**2-Methylcyclopentanone**	**1400**	**Derived EU-LCI (read-across)**	**2020**
8-7	583-60-8	**2-Methylcyclohexanone**	**2300**	**Ascribed EU-LCI**	**2013**
8-8	98-86-2	**Acetophenone**	**490**	**Ascribed EU-LCI**	**2013**
8-9	116-09-6	**1-Hydroxyacetone (1-hydroxy-2- propanone)**	**2100**	**Derived EU-LCI (read-across)**	**2019**
8-10	67-64-1	**Acetone**	**120000**	**Derived EU-LCI**	**2018**
9		***Acids***			
9-1	64-19-7	**Acetic acid**	**1200**	**Derived EU-LCI**	**2016**
9-2	79-09-4	**Propionic acid**	**1500**	**Derived EU-LCI**	**2016**
9-3	79-31-2	**Isobutanoic acid (isobutyric acid)**	**1800**	**Derived EU-LCI (read-across)**	**2018**
9-4	107-92-6	**Butanoic acid (butyric acid)**	**1800**	**Derived EU-LCI (read-across)**	**2018**
9-5	75-98-9	**2,2-Dimethylpropanoic acid (pivalic acid)**	**2100**	**Derived EU-LCI (read-across)**	**2018**
9-6	109-52-4	**n-Pentanoic acid (valeric acid)**	**2100**	**Derived EU-LCI (read-across)**	**2018**
9-7	142-62-1	**n-Hexanoic acid (caproic acid)**	**2100**	**Derived EU-LCI (read-across)**	**2018**
9-8	111-14-8	**n-Heptanoic acid**	**2100**	**Derived EU-LCI (read-across)**	**2018**

No.	*CAS no.*	*Compound*	*EU-LCI ($\mu g/m^3$)*	*Status of EU-LCI value*	*Year of adoption*
9-9	124-07-2	**n-Octanoic acid**	**2100**	**Derived EU-LCI (read-across)**	**2018**
9-10	149-57-5	**2-Ethylhexanoic acid**	**150**	**Derived EU-LCI**	**2014**
10		***Ester***			
10-1	108-21-4	**Propyl acetate (n-, iso-)**	**4200**	**Ascribed EU-LCI**	**2013**
10-2	108-65-6	**2-Methoxy-1-methylethyl acetate**	**650**	**Derived EU-LCI**	**2019**
10-3*	107-31-3	**Methyl formate**	**3000**	**Derived EU-LCI**	**2021**
10-4*	592-84-7	**n-Butyl formate**	**4900**	**Derived EU-LCI**	**2021**
10-5	80-62-6	**Methyl methacrylate**	**750**	**Derived EU-LCI**	**2016**
10-7	110-19-0	**Isobutyl acetate**	**4800**	**Ascribed EU-LCI**	**2013**
10-8	123-86-4	**n-Butyl acetate**	**4800**	**Ascribed EU-LCI**	**2013**
10-9	103-09-3	**2-Ethylhexyl acetate**	**350**	**Derived EU-LCI (read-across)**	**2018**
10-10	96-33-3	**Methyl acrylate**	**180**	**Ascribed EU-LCI**	**2013**
10-11	140-88-5	**Ethyl acrylate**	**200**	**Ascribed EU-LCI**	**2013**
10-12	141-32-2	**n-Butyl acrylate**	**110**	**Ascribed EU-LCI**	**2013**
10-13	103-11-7	**2-Ethylhexyl acrylate**	**380**	**Ascribed EU-LCI**	**2013**
10-14		**Other acrylates (acrylic acid esters)**	**110**	**Ascribed EU-LCI**	**2013**
10-15	627-93-0	**Dimethyl adipate**	**50**	**Ascribed EU-LCI**	**2013**
10-16	106-65-0	**Dimethyl succinate**	**20**	**Derived EU-LCI**	**2020**
10-17	1119-40-0	**Dimethyl glutarate**	**25**	**Derived EU-LCI**	**2020**
10-18	71195-64-7	**Diisobutyl glutarate**	**35**	**Derived EU-LCI (read-across)**	**2020**
10-19	925-06-4	**Diisobutyl succinate**	**35**	**Derived EU-LCI (read-across)**	**2020**
10-20	105-75-9	**Dibutyl fumarate**	**50**	**Ascribed EU-LCI**	**2013**
10-21	105-76-0	**Maleic acid dibutylester**	**50**	**Ascribed EU-LCI**	**2013**
10-22	13048-33-4	**Hexamethylene diacrylate**	**10**	**Ascribed EU-LCI**	**2013**
10-23	96-48-0	**Butyrolactone**	**2800**	**Derived EU-LCI**	**2018**
11		***Chlorinated hydrocarbons***			
11-1	127-18-4	**Tetrachloroethene**	**80**	**Derived EU-LCI**	**2018**
11-3	106-46-7	**1,4-Dichlorobenzene**	**150**	**Derived EU-LCI**	**2013**
12		***Others***			
12-1	123-91-1	**1,4-Dioxane**	**400**	**Derived EU-LCI**	**2015**
12-2	105-60-2	**Caprolactame**	**300**	**Derived EU-LCI**	**2013**
12-3	872-50-4	**N-Methyl-2-pyrrolidone**	**1800**	**Derived EU-LCI**	**2016**
12-4	556-67-2	**Octamethylcyclotetrasiloxane (D4)**	**1200**	**Ascribed EU-LCI**	**2013**
12-7	100-97-0	**Hexamethylenetetramine**	**30**	**Ascribed EU-LCI**	**2013**
12-8	96-29-7	**2-Butanonoxime**	**15**	**Derived EU-LCI**	**2015**
12-9	126-73-8	**Tributyl phosphate**	**300**	**Derived EU-LCI**	**2016**
12-10*	78-40-0	**Triethyl phosphate**	**1700**	**Derived EU-LCI**	**2021**
12-11	26172-55-4	**5-Chloro-2-methyl-2H-isothiazol-3-one (CIT)**	**1**	**Ascribed EU-LCI**	**2013**
12-12	2682-20-4	**2-Methyl-4-isothiazolin-3-one (MIT)**	**100**	**Ascribed EU-LCI**	**2013**
12-13	121-44-8	**Triethylamine**	**60**	**Derived EU-LCI**	**2017**
12-14	109-99-9	**Tetrahydrofuran**	**500**	**Derived EU-LCI**	**2018**
12-17	2687-91-4	**N-Ethyl-2-pyrrolidone**	**400**	**Derived EU-LCI**	**2016**

* new or altered

Index

Printed in the United States
by Baker & Taylor Publisher Services